Advanced Maths Essentials
Core 2 for AQA

Welcome to Advanced Maths Essentials: Co
your examination performance by focusing o
your AQA Core 2 examination. It has been d
to be studied. Each chapter has then been div
each sub-heading gives the AQA specificatio

The book contains scores of worked example
problem. You can then apply the steps to solv
exam questions at the end of each chapter. If you feel you need extra practice on any topic,
you can try the Skills Check Extra exercises on the accompanying CD-ROM. At the back of
this book there is a sample exam-style paper to help you test yourself before the big day.

Some of the questions in the book have a  symbol next to them. These questions have a
PowerPoint® solution (on the CD-ROM) that guides you through suggested steps in solving
the problem and setting out your answer clearly.

Using the CD-ROM

To use the accompanying CD-ROM simply put the disc in your CD-ROM drive, and the menu
should appear automatically. If it doesn't automatically run on your PC:

1. Select the My Computer icon on your desktop.
2. Select the CD-ROM drive icon.
3. Select Open.
4. Select core2_for _aqa.exe.

If you don't have PowerPoint® on your computer you can download PowerPoint 2003
Viewer®. This will allow you to view and print the presentations. Download the viewer from
http://www.microsoft.com

Pearson Education Limited
Edinburgh Gate
Harlow
Essex
CM20 2JE
England
www.longman.co.uk

First published 2005
Fourth impression 2010
ISBN   978-0-582-83680-8

Design by Ken Vail Graphic Design

Cover design by Raven Design

Typeset by Tech-Set, Gateshead

Printed in Malaysia, CTP-VVP

The publisher's policy is to use paper manufactured from sustainable forests.

The Publisher wishes to draw attention to the Single-User Licence Agreement situated at the back of the book. Please read this agreement carefully before installing and using the CD-ROM.

We are grateful for permission from the Assessment and Qualifications Alliance to reproduce past exam questions. All such questions have a reference in the margin. The Assessment and Qualifications Alliance can accept no responsibility whatsoever for accuracy of any solutions or answers to these questions.

Every effort has been made to ensure that the structure and level of sample question papers matches the current specification requirements and that solutions are accurate. However, the publisher can accept no responsibility whatsoever for accuracy of any solutions or answers to these questions. Any such solutions or answers may not necessarily constitute all possible solutions.

1 Algebra and functions

1.1 Indices

Laws of indices for all rational exponents.

The laws or rules of indices allow you to simplify terms that are written in **index form**, a^m, where m is rational.

a is the **base**, where $a \neq 0$.

m is the **index**, also known as the **power** or **exponent**.

Note:
The index m can be a positive or negative integer, or a fraction, or zero.

These rules apply to terms in index form with the same base:

> ***Rule 1*** To **multiply** the terms, **add** the indices:
> $$a^m \times a^n = a^{m+n}$$
>
> ***Rule 2*** To **divide** the terms, **subtract** the indices:
> $$a^m \div a^n = a^{m-n}$$
>
> ***Rule 3*** To **raise to a power**, **multiply** the indices:
> $$(a^m)^n = a^{mn}$$

Tip:
$a^m \div a^n$ is also written $\dfrac{a^m}{a^n}$.

Example 1.1 Simplify   **a** $2x^5 \times 7x^6$ **b** $10x^5z^3 \div 2x^3z$ **c** $(2p^2)^5$.

Step 1: Gather like terms and simplify using the index laws.

a $2x^5 \times 7x^6 = 2 \times 7 \times x^5 \times x^6$
$= 14 \times x^{5+6}$ (*Rule 1*)
$= 14x^{11}$

Tip:
Multiply or divide the numbers, then deal with the expressions in index form.

b $10x^5z^3 \div 2x^3z = \dfrac{10x^5z^3}{2x^3z}$

$= 5x^{5-3}z^{3-1}$ (*Rule 2*)

$= 5x^2z^2$

Tip:
When no index is written, this means the power is 1, so $z = z^1$.

c $(2p^2)^5 = 2^5(p^2)^5 = 32p^{10}$ (*Rule 3*)

Tip:
$(ab)^n = a^n b^n$, so remember to raise 2 to the power 5 here.

Example 1.2 **a** Write each of these expressions as a power of 2:

   i 8^4 **ii** 4^{x+1}

b Hence solve the equation $8^4 = 4^{x+1}$.

Step 1: Write each term in index form with the same base and simplify, using the index laws.

a **i** $8^4 = (2^3)^4 = 2^{12}$ (*Rule 3*)
   ii $4^{x+1} = (2^2)^{x+1} = 2^{2(x+1)} = 2^{2x+2}$ (*Rule 3*)

Step 2: Equate the indices and solve.

b $8^4 = 4^{x+1} \Rightarrow 2^{12} = 2^{2x+2}$

Equating indices gives $12 = 2x + 2$
$$x = 5$$

The zero index, a^0

You know that $a^n \div a^n = 1$

But, by Rule 3, $a^n \div a^n = a^{n-n} = a^0$

$\Rightarrow$ $a^0 = 1$ ***Rule 4***

Note:
a cannot be zero; 0^0 is undefined.

Negative index, a^{-n}

You know that $\qquad a^n \times a^{-n} = a^0 = 1$

Divide both sides by a^n $\quad a^{-n} = \dfrac{1}{a^n}$ **Rule 5a**

This format is useful when working in fractions:

$$\left(\dfrac{a}{b}\right)^{-n} = \left(\dfrac{b}{a}\right)^n$$ **Rule 5b**

Fractional indices

$$a^{\frac{1}{n}} = \sqrt[n]{a}$$ **Rule 6a**

For example, $a^{\frac{1}{3}} = \sqrt[3]{a}$.

$$a^{\frac{m}{n}} = (\sqrt[n]{a})^m = \sqrt[n]{a^m}$$ **Rule 6b**

For example, $a^{\frac{2}{3}} = (\sqrt[3]{a})^2 = \sqrt[3]{(a^2)}$.

To calculate $64^{\frac{2}{3}}$ you could find $(\sqrt[3]{64})^2 = 4^2 = 16$.

Alternatively, you could find $\sqrt[3]{(64^2)} = \sqrt[3]{4096} = 16$.

Example 1.3 Evaluate, without using a calculator,

    a $3^4 \div 3^7$ **b** $\left(\tfrac{3}{4}\right)^{-1}$ **c** $4^{\frac{1}{2}}$ **d** $8^{-\frac{1}{3}}$ **e** $\left(\tfrac{1}{125}\right)^{-\frac{2}{3}}$

Step 1: Use the index laws to calculate the values.

a $3^4 \div 3^7 = 3^{-3} = \dfrac{1}{3^3} = \dfrac{1}{27}$ *(Rules 2 & 5a)*

b $\left(\tfrac{3}{4}\right)^{-1} = \left(\tfrac{4}{3}\right)^1 = \tfrac{4}{3}$ *(Rule 5b)*

c $4^{\frac{1}{2}} = \sqrt{4} = 2$ *(Rule 6a)*

d $8^{-\frac{1}{3}} = \dfrac{1}{8^{\frac{1}{3}}} = \dfrac{1}{\sqrt[3]{8}} = \dfrac{1}{2}$ *(Rules 5a & 6a)*

e $\left(\tfrac{1}{125}\right)^{-\frac{2}{3}} = 125^{\frac{2}{3}} = (\sqrt[3]{125})^2 = 5^2 = 25$ *(Rules 5b & 6b)*

Example 1.4 Given that $f(x) = 6x^3 + x$ and $g(x) = \sqrt{x}$, express, in index form,

    a $f(x) \times g(x)$ **b** $f(x) \div g(x)$

Step 1: Expand the brackets.

a $f(x) \times g(x) = (6x^3 + x) \times \sqrt{x}$

Step 2: Simplify using the index laws.

$\qquad\qquad\quad = 6x^3 \times x^{\frac{1}{2}} + x^1 \times x^{\frac{1}{2}}$

$\qquad\qquad\quad = 6x^{\frac{7}{2}} + x^{\frac{3}{2}}$ *(Rule 1)*

Step 1: Divide each term of the numerator by the denominator.

b $f(x) \div g(x) = \dfrac{6x^3 + x}{\sqrt{x}}$

$\qquad\qquad\quad = \dfrac{6x^3}{x^{\frac{1}{2}}} + \dfrac{x}{x^{\frac{1}{2}}}$

Step 2: Simplify using the index laws.

$\qquad\qquad\quad = 6x^{\frac{5}{2}} + x^{\frac{1}{2}}$ *(Rule 2)*

1 Simplify:

   a $2x^3y^5 \times 3xy^{-1}$ **b** $(2a^2)^4$ **c** $14pq^7 \div 2p^2q^5$

2 Evaluate, without using a calculator:

   a $\dfrac{1}{2^{-1}}$ **b** 3^{-2} **c** $27^{-\frac{1}{3}}$

   d $16^{\frac{3}{2}}$ **e** $0.25^{-\frac{1}{2}}$

3 Simplify $3a^2b^{-2} \times 4a^3 \sqrt{b}$.

4 Write as a single power of x:

   a $x^2\sqrt{x}$ **b** $\dfrac{\sqrt{x}(\sqrt{x})^3}{x^3}$ **c** $\dfrac{\sqrt{x}(\sqrt{x})^3}{x^{-3}}$

5 Write $\dfrac{p^{\frac{1}{6}}p^{\frac{2}{3}}}{\sqrt{p}}$ in the form p^k where k is a number to be found.

6 **a** Write each of the following as a power of 2:   **i** 4^x **ii** 8^{x-1}.

   b Express 4^x8^{x-1} as a single power of 2.

   c Solve the equation $4^x = 8^{x-1}$.

7 **a** Express 9^{2x} as a power of 3.

   b Solve $3^{x-1} = 9^{2x}$.

   c Solve $9^{2x} = \dfrac{1}{3^3}$.

8 **a** Given that $8^{2x-1} = 4^y$, form an equation in the form $y = ax + b$, where a and b are rational numbers to be found.

   b Given also that $9^{x+1} = \dfrac{81^{y-1}}{27}$, form another equation relating x and y.

   c Hence find the values of x and y.

9 Given that $7^{x+6} = 49^{2x}$, find x.

10 **a** Express 36^{2p} as a power of 6.

   b Express 216^{q-1} as a power of 6.

   c Given that $36^{2p} = 216^{q-1}$, form a linear equation in p and q.

   d Given also that $p = 3q$, find the values of p and q.

SKILLS CHECK **1A EXTRA** is on the CD

Knowledge of the effect of simple transformations on the graph of $y = f(x)$ as represented by $y = af(x), y = f(x) + a, y = f(x + a), y = f(ax)$.

In module C1 you studied translations of a parabola and a circle (C1 Section 1.15). In C2, the work on translations is extended to other curves. You also need to know about reflections in the x- and y-axes and stretches in the x- and y-directions.

The translations and stretches are illustrated below, using, as $y = f(x)$, the graph of $y = \sin x \ (0° \leqslant x \leqslant 360°)$.

Note:
See Section 3.5 for more about the graph of $y = \sin x$.

Translations

Recall:
Translations, C1 Section 1.15.

$y = f(x) + a$

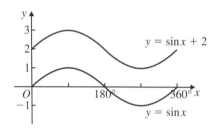

In the graph of $y = \sin x + 2$, all the points in $y = \sin x$ have moved up two units. This is a translation.

In general, $y = f(x) + a$ is a translation a units in the y-direction of $y = f(x)$.

The vector form of the translation is $\begin{bmatrix} 0 \\ a \end{bmatrix}$.

If $a > 0$, the graph moves up.

If $a < 0$, the graph moves down.

$y = f(x + a)$

Recall:
Translations, C1 Section 1.15.

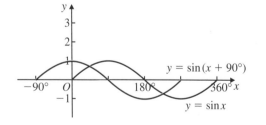

Note:
The new graph is $y = \cos x$ (see Section 3.5).

In the graph of $y = \sin(x + 90°)$, all the points in $y = \sin x$ have moved 90° to the left. This is a translation.

In general, $y = f(x + a)$ is a translation $-a$ units in the x-direction of $y = f(x)$. The vector form of the translation is $\begin{bmatrix} -a \\ 0 \end{bmatrix}$.

If $a > 0$, the graph moves to the left.

If $a < 0$, the graph moves to the right.

Stretches

$y = af(x)$

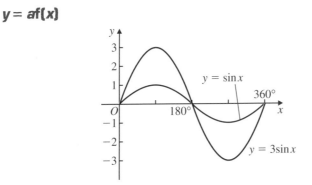

In the graph of $y = 3\sin x$, all the points in $y = \sin x$ have been stretched in the y-direction by factor 3.

In general, $y = af(x)$ is a stretch in the y-direction, by factor a units, of $y = f(x)$. Points on the x-axis are invariant.

$y = f(ax)$

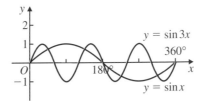

In the graph of $y = \sin 3x$, all the points have been stretched in the x-direction, by factor $\frac{1}{3}$.

In general, $y = f(ax)$ is a stretch in the x-direction by factor $\dfrac{1}{a}$ units, of $y = f(x)$. Points on the y-axis are invariant.

When $a > 1$, the graph appears more squashed in the x-direction.

For $0 < a < 1$, the graph appears to be lengthened in the x-direction.

In both cases, the transformation is described as a stretch.

Reflections

Reflections in the x-axis and the y-axis are illustrated below using, as $y = f(x)$, the line $y = x + 2$.

$y = -f(x)$

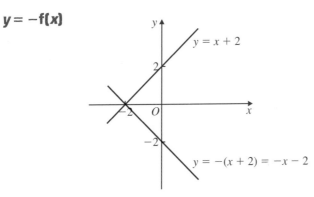

In the graph of $y = -(x + 2)$, all the points in the graph of $y = x + 2$ have been reflected in the x-axis.

In general, $y = -f(x)$ is a reflection in the x-axis of $y = f(x)$.

Tip:
All the y-coordinates are multiplied by 3, so the vertex (90°, 1) moves to (90°, 3).

Note:
Invariant points do not move under the transformation.

Tip:
All the x-coordinates are divided by 3, so the vertex (90°, 1) moves to (30°, 1).

Note:
This is called a stretch, even though it looks squashed up.

Tip:
In the examination you will not gain the marks unless you use the correct terminology of **stretch** in the x-direction. Remember to include the scale factor.

$y = f(-x)$

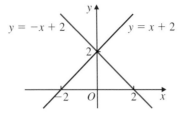

In the graph of $y = -x + 2$, all the points in the graph of $y = x + 2$ have been reflected in the y-axis.

In general, $y = f(-x)$ is a reflection in the y-axis of $y = f(x)$.

Mixed examples

Example 1.5 The diagram shows a sketch of $y = x(2 - x)$.

The vertex P is the point $(1, 1)$.

By applying appropriate transformations, sketch the following curves:

a $y = 3x(2 - x)$ **b** $y = x(2 - x) - 1$

In each case, describe the transformation and state the coordinates of P_1, the vertex of the curve.

Step 1: Identify the transformation.

Step 2: Sketch the curve, applying the transformation.

Step 3: Describe the transformation and state the coordinates of the vertex.

a

$y = 3x(2 - x)$ is a stretch in the y-direction, factor 3 of the curve $y = x(2 - x)$.

The coordinates of the vertex P_1 are $(1, 3)$.

Tip:
The y-coordinates are multiplied by 3.

b

$y = x(2 - x) - 1$ is a translation by $\begin{bmatrix} 0 \\ -1 \end{bmatrix}$ of the curve $y = x(2 - x)$.

The coordinates of the vertex P_1 are $(1, 0)$.

Tip:
All points move 1 unit down.

Example 1.6 The diagram shows the curves $y = 3^x$ and $y = 3^{x+2}$.

Note:
See Section 4.1 for the graph of $y = a^x$.

a Describe the translation that maps $y = 3^x$ onto $y = 3^{x+2}$.

b Describe another transformation that maps $y = 3^x$ onto $y = 3^{x+2}$.

Step 1: Identify the translation $y = f(x + a)$.

a The *translation* that maps $y = 3^x$ onto $y = 3^{x+2}$ is one that moves the curve 2 units to the left, that is, by the vector $\begin{bmatrix} -2 \\ 0 \end{bmatrix}$.

Step 2: Rearrange into the form $y = af(x)$ and identify the alternative transformation.

b $y = 3^{x+2} = 3^x \times 3^2 = 9 \times 3^x$

To map $y = 3^x$ onto $y = 3^{x+2}$, apply a *stretch* in the y-direction, factor 9.

Recall:
Index laws (Section 1.1).

Example 1.7 The diagram shows a sketch of $y = 2^x$.

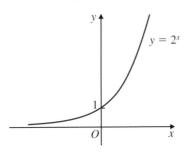

State the transformation that gives the curve $y = 2^{-x}$ and sketch $y = 2^{-x}$.

Step 1: Identify the transformation $y = f(-x)$.

To map $y = 2^x$ onto $y = 2^{-x}$, reflect in the y-axis.

Step 2: Sketch the curve.

Example 1.8 The diagram shows a sketch of $y = f(x)$ for $0 \leqslant x \leqslant 3$. For all other values of x, $f(x) = 0$.

A is the point $(2, 1)$ and B is the point $(3, 0)$.

Draw sketches to show

a $y = f(2x)$ **b** $y = f(\frac{1}{3}x)$

Describe the transformations, stating the new positions of A and B.

Step 1: Identify the appropriate transformation.

Step 2: Draw the sketch.

Step 3: Describe the transformation and state the new coordinates.

a

This is a stretch in the x-direction, factor $\frac{1}{2}$.
A moves to $(1, 1)$ and B moves to $(1.5, 0)$.

Tip:
In **a**, x-coordinates are halved, maintaining the y-coordinates.

b

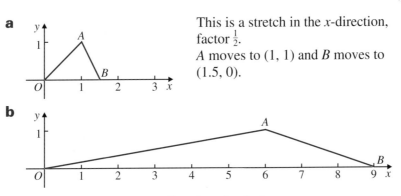

This is a stretch in the x-direction, factor 3.
A moves to $(6, 1)$ and B moves to $(9, 0)$.

Tip:
In **b**, the factor of stretch $= \dfrac{1}{\frac{1}{3}} = 3$

Tip:
In **b**, x-coordinates are multiplied by 3, maintaining the y-coordinates.

See Chapter 3 for more on trigonometric curves.
See Chapter 4 for more on exponentials and logarithms.

1 a Describe the transformation that maps $y = x^2$ onto   **i** $y = x^2 + 2$ **ii** $y = -x^2$.

  b Sketch the three curves on the same axes.

2 On the same axes, sketch $y = \cos x$ and $y = \cos(x + 30°)$ where $0° \le x \le 360°$.

3 Describe the transformation that maps $y = \cos x$ onto

  a $y = 1 + \cos x$ **b** $y = 4\cos x$ **c** $y = -\cos x$ **d** $y = \cos 2x$

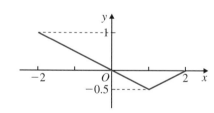

4 The diagram shows the graph $y = f(x)$
for $0 \le x \le 2$. The vertex A is the point $(1, 3)$.

  a Sketch the graph of $y = -f(x)$ for $0 \le x \le 2$ and state
  the coordinates of A_1, the vertex of the curve $y = -f(x)$.

  b i Describe fully the transformation that maps
    $y = f(x)$ onto $y = 2f(x)$.

   ii Sketch the graph of $y = 2f(x)$, for $0 \le x \le 2$
    and state the coordinates of A_2, the vertex of the curve $y = 2f(x)$.

5 On the same axes, sketch $y = \sin x$ and $y = \sin(x - \frac{1}{4}\pi)$ for $0 \le x \le 2\pi$.

6 a Given that $f(x) = x^2$, state the transformation that maps $y = f(x)$ onto
   i $y = 4f(x)$ **ii** $y = f(2x)$

  b State the equation of each of the transformed curves and comment.

7 The sketch of $y = f(x)$ for $-2 \le x \le 2$ is shown
in the diagram.

On separate axes, for $-2 \le x \le 2$, sketch the graphs of

  a $y = f(-x)$

  b $y = -f(x)$

  c $y = f(x) - 1$

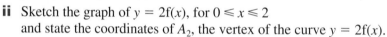

8 In each of the following parts, sketch the two graphs on the same set of axes, showing any intercepts
with the axes:

  a $y = 2^x$ and $y = 2^x + 2$ **b** $y = 2^x$ and $y = -2^x$ **c** $y = 2^x$ and $y = 2^{x-2}$

9 Which of these curves is the same as $y = \cos x$?

  a $y = \sin(x - 90°)$ **b** $y = \cos(-x)$ **c** $y = -\cos x$ **d** $y = \cos(x + 360°)$

10 Describe the transformation that maps $y = \tan x$ onto $y = \tan 2x$ and sketch $y = \tan 2x$ for
$0° \le x \le 180°$.

1 **a** Write each of the following as a power of 3: **i** $\frac{1}{27}$ **ii** 9^x.

   b Hence solve the equation $9^x \times 3^{1-x} = \frac{1}{27}$. [AQA (B) May 2002]

2 **a** Express each of the following as a power of 3: **i** $\sqrt{3}$ **ii** $\dfrac{3^x}{\sqrt{3}}$.

   b Hence, or otherwise, solve the equation $\dfrac{3^x}{\sqrt{3}} = \dfrac{1}{3}$. [AQA (A) May 2003]

3 **i** Write $\sqrt{2}$ as a power of 2.

   ii Hence express $4\sqrt{2}$ as a power of 2.

   iii Hence solve the equation $2^{3x+4} = 4\sqrt{2}$. [AQA (B) Nov 2002]

4 Given that $p = x^{\frac{3}{2}}$ and $q = x^{-\frac{5}{2}}$, express, in index form in terms of x:

   a pq **b** $p \div q$ **c** $\sqrt{p}$

5 Simplify $(2x^{\frac{1}{2}})^4$.

6 **a** Given that $f(x) = \sin x$, sketch the graph of $y = f(x)$ for $-\pi \leqslant x \leqslant 2\pi$.

   b On the same axes, sketch $y = f(x - \frac{1}{2}\pi)$.

7 The diagram shows a sketch of the graph $y = \cos 2x$ with a line of symmetry L.

   i Describe the geometrical transformation by which the graph of $y = \cos 2x$ can be obtained from that of $y = \cos x$.

   ii Write down the equation of the line L. [AQA (A) Jan 2002]

8 The graph of $y = \log_{10} x$ is mapped to the graph of $y = \log_{10}(100x)$.

   a Write $\log_{10}(100x)$ in the form $a + \log_{10} x$, where a is an integer, and hence state the vector of the translation that describes the mapping.

> **Hint:** Laws of logarithms, page 48.

   b Given that the mapping is a stretch, state the scale factor and direction of the stretch.

9 **a** Describe the translation that maps the curve $y = 2^x$ onto the curve $y = 2^{x-1}$.

   b Describe another transformation that maps the curve $y = 2^x$ onto the curve $y = 2^{x-1}$.

10 The diagram shows a sketch of $y = f(x)$.
A is the point $(-4, 0)$, B is $(-2, 2)$, C is $(0, 1)$ and D is $(2, 0)$.

State the coordinates of the new positions of A, B, C and D when $y = f(x)$ is mapped onto

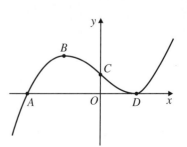

   a $y = 2f(x)$

   b $y = f(\frac{1}{2}x)$.

Sequences and series

2.1 Introducing sequences and series

Sequences, including those given by a formula for the *n*th term.

A **sequence** is a succession of terms that follow a rule. One way of defining a sequence is to give the formula for the *n*th term. The sequence is generated by substituting positive integer values of *n* into the formula.

> **Note:**
> The *n*th term is often denoted by u_n, x_n or t_n.

Example 2.1 The *n*th term of a sequence is u_n, where $u_n = 2n - 3$ for $n \geq 1$. Write down the first four terms of the sequence.

Step 1: Substitute $n = 1, 2, 3$ and 4 into the formula.

$n = 1 \qquad u_1 = 2 \times 1 - 3 = -1$

$n = 2 \qquad u_2 = 2 \times 2 - 3 = 1$

$n = 3 \qquad u_3 = 2 \times 3 - 3 = 3$

$n = 4 \qquad u_4 = 2 \times 4 - 3 = 5$

> **Note:**
> This is an example of an arithmetic sequence. See Section 2.3.

The first four terms of the sequence are $-1, 1, 3, 5$.

Convergent sequences

A sequence is said to **converge** if, as *n* gets larger, the terms of the sequence get closer and closer to a particular number, called the **limiting value**.

Example 2.2 **a** Find the first five terms of the sequence defined by
$u_n = 3 + \left(\frac{1}{2}\right)^n$, $n \geq 1$.

b The sequence converges to *L*. Find *L*.

Step 1: Substitute $n = 1, 2, 3, 4$ and 5 into the formula.

a $n = 1 \qquad u_1 = 3 + \left(\frac{1}{2}\right)^1 = 3.5$

$n = 2 \qquad u_2 = 3 + \left(\frac{1}{2}\right)^2 = 3.25$

$n = 3 \qquad u_3 = 3 + \left(\frac{1}{2}\right)^3 = 3.125$

$n = 4 \qquad u_4 = 3 + \left(\frac{1}{2}\right)^4 = 3.0625$

$n = 5 \qquad u_5 = 3 + \left(\frac{1}{2}\right)^5 = 3.03125$

> **Tip:**
> For large values of *n*, $\left(\frac{1}{2}\right)^n$ gets closer and closer to 0.

The sequence is $3.5, 3.25, 3.125, 3.0625, 3.03125, \ldots$

Step 2: Consider what happens when *n* is large.

b As $n \to \infty$, $\left(\frac{1}{2}\right)^n \to 0$, hence $3 + \left(\frac{1}{2}\right)^n \to 3$.

Therefore $L = 3$.

> **Tip:**
> The symbol $\to$ means 'tends to'.

2.2 Recurrence relations

Sequences generated by a simple relation in the form $x_{n+1} = f(x_n)$.

Another way of defining a sequence is to give a **recurrence relation**. This could be in the form $x_{n+1} = f(x_n)$, where a subsequent term is given in relation to the previous term. To generate the sequence, you must know the first term, x_1.

> **Note:**
> This is also known as an **iterative formula**.

Example 2.3 A sequence is defined by the recurrence relation $x_{n+1} = 3x_n - 2$, with $x_1 = 4$. Calculate x_2, x_3 and x_4.

Step 1: Substitute $n = 1, 2$ and 3 into the formula.

$n = 1 \qquad x_2 = 3x_1 - 2 = 3 \times 4 - 2 = 10$

$n = 2 \qquad x_3 = 3x_2 - 2 = 3 \times 10 - 2 = 28$

$n = 3 \qquad x_4 = 3x_3 - 2 = 3 \times 28 - 2 = 82$

> **Note:**
> In this sequence the terms do not appear to converge.

Calculator note:

If your calculator has an ANS key, try generating the terms as follows:

		Display
Enter the first value	4 =	4
Write the formula	3 × ANS − 2 =	10
Generate next term	=	28
Generate next term	=	82

Tip:
You may not have to press ×.

Tip:
The next term is generated every time you press =.
Keep on pressing!

Example 2.4 The value of a property is estimated to increase each year, where its price in each subsequent year is related to the current year by the recurrence relation $x_{n+1} = 1.05x_n$.

Given that the value of the property at the end of 2003 was £120 000, find the estimated value of the property at the end of 2007, to the nearest £1000.

Note:
This recurrence relation is describing a 5% increase in value each year.

Step 1: Substitute $n = 1, 2, 3, \ldots$ successively into the formula.

Value in 2003	$x_1 = 120\,000$
Value in 2004	$x_2 = 1.05x_1 = 1.05 \times 120\,000 = 126\,000$
Value in 2005	$x_3 = 1.05x_2 = 1.05 \times 126\,000 = 132\,300$
Value in 2006	$x_4 = 1.05x_3 = 1.05 \times 132\,300 = 138\,915$
Value in 2007	$x_5 = 1.05x_4 = 1.05 \times 138\,915 = 145\,860.75$

Note:
This is an example of a geometric sequence. See Section 2.4.

The estimated value of the property at the end of 2007, to the nearest £1000, is £146 000.

Finding the limit of a convergent sequence

If the terms of the recurrence relation $x_{n+1} = f(x_n)$ converge, then you can find the limit, L, of the sequence by solving the equation $L = f(L)$.

Tip:
Replace both x_{n+1} and x_n by L.

Example 2.5 A sequence has recurrence relation $x_{n+1} = \dfrac{4}{x_n - 1} + 4$, $x_1 = 1.5$.

a Calculate x_2, x_3, x_4, x_5, x_6 and x_7.

b The terms approach a limit L. Suggest the value of L.

c Show that L satisfies the equation $L^2 - 5L = 0$ and hence calculate the value of L.

Step 1: Substitute $n = 1, 2, 3, \ldots$ successively into the formula.

a $x_1 = 1.5$

$$x_2 = \frac{4}{1.5 - 1} + 4 = 12$$

$$x_3 = \frac{4}{12 - 1} + 4 = 4.363636\ldots$$

$$x_4 = \frac{4}{4.363636\ldots - 1} + 4 = 5.189189\ldots$$

$$x_5 = \frac{4}{5.189189\ldots - 1} + 4 = 4.9548387\ldots$$

$$x_6 = \frac{4}{4.9548387\ldots - 1} + 4 = 5.0114192\ldots$$

$$x_7 = \frac{4}{5.0114192\ldots - 1} + 4 = 4.9971533\ldots$$

Tip:
Use all the digits on your calculator in each stage of your calculation.
Alternatively, generate the terms on your calculator thus:

1.5 =

4 ÷ ((ANS − 1)) + 4

=

=

= etc.

Step 2: Make a guess at the limit.

b It appears that the sequence converges to 5.

Step 3: Solve $L = f(L)$. **c**

$$L = \frac{4}{L-1} + 4$$

$$L(L-1) = 4 + 4(L-1)$$

$$L^2 - L = 4 + 4L - 4$$

$$L^2 - 5L = 0$$

$$L(L-5) = 0$$

$$\Rightarrow L = 0, L = 5$$

Tip:
Factorise and solve.

Step 4: Compare with known values in the sequence.

Therefore the limit of the sequence is 5.

Tip:
Choose the appropriate value.

Series

When the terms of a sequence are added, a **series** is formed, for example $2 + 6 + 10 + 14 + 18 + 22$.

There is a shorthand way of writing the terms of a series, using a general term and Σ (sigma) notation.

Note:
Σ means 'the sum of' and is read 'sigma'.

For example, to evaluate $\sum_{r=1}^{5} (2r + 3)$, add the terms calculated by substituting $r = 1, 2, 3, 4$ and 5.

When $r = 1$, $2r + 3 = 5$; when $r = 2$, $2r + 3 = 7$, and so on.

So $\sum_{r=1}^{5} (2r + 3) = 5 + 7 + 9 + 11 + 13 = 45$.

Example 2.6 Evaluate $\sum_{r=1}^{4} u_r$ where **a** $u_r = 6 - 3r$ **b** $u_r = 3(2^r)$.

Note:
Letters other than r may be used, for example i.

Step 1: Substitute $r = 1, 2, 3, 4$.

a $u_1 = 6 - 3 \times 1 = 3$ $\qquad u_2 = 6 - 3 \times 2 = 0$

$u_3 = 6 - 3 \times 3 = -3$ $\qquad u_4 = 6 - 3 \times 4 = -6$

Note:
This is an arithmetic series (Section 2.3).

Step 2: Add the terms.

$$\sum_{r=1}^{4} u_r = 3 + 0 + -3 + -6 = -6$$

Step 1: Substitute $r = 1, 2, 3, 4$.

b $u_1 = 3(2^1) = 6$

$u_2 = 3(2^2) = 12$

$u_3 = 3(2^3) = 24$

$u_4 = 3(2^4) = 48$

Note:
This is a geometric series (Section 2.4).

Step 2: Add the terms.

$$\sum_{r=1}^{4} u_r = 6 + 12 + 24 + 48 = 90$$

SKILLS CHECK **2A: Introducing sequences and series; recurrence relations**

1 The nth term, u_n, of a sequence is $n^2 - 1$. Find the first four terms.

2 a If $x_n = 2 - \left(\frac{2}{3}\right)^n$, find x_1, x_2, x_3 and x_4.

b The sequence converges to L. Find L.

3 Calculate the next four terms of the sequence defined by the recurrence relation

    a $x_{n+1} = 2x_n$, $x_1 = 2$
        b $x_{n+1} = 2x_n + 1$, $x_1 = 3$

4 A car cost £10 000 when new. It depreciated in value by 10% each year.

    a Write down a recurrence relation for the value of the car from one year to the next.

    b Using the recurrence relation, find the value of the car five years from when it was bought, giving your answer to the nearest £100.

5 A sequence of numbers is defined by the recurrence relation $v_{n+1} = \sqrt{v_n}$, $v_1 = 100$.

    a Write down the next seven terms of the sequence, giving your answer correct to four decimal places where appropriate.

    b The sequence converges to L. Show algebraically that $L = 1$.

6 A sequence has recurrence relation $x_{n+1} = \dfrac{5 - 4x_n}{x_n}$, $x_1 = -6$.

    a Calculate x_2, x_3, x_4, x_5, x_6 and x_7.

    b The terms approach a limit L. Suggest the value of L.

    c Show that L satisfies the equation $L^2 + 4L - 5 = 0$ and hence calculate the value of L.

7 Evaluate $\displaystyle\sum_{r=1}^{5} u_r$ where

    a $u_r = 3 + 4r$
        b $u_r = 0.5(2^r)$

8 Evaluate

    a $\displaystyle\sum_{r=1}^{3} (20 - 3r)$
        b $\displaystyle\sum_{r=1}^{4} 2^{r-1}$

9 A recurrence relation is defined by $u_{r+2} = 2u_{r+1} - 3u_r - 1$, where $u_1 = 1$, $u_2 = 0$. Find u_3, u_4 and u_5.

10 Evaluate

    a $\displaystyle\sum_{r=1}^{5} (r^2 - r)$
        b $\displaystyle\sum_{r=4}^{6} \dfrac{r}{r+1}$

SKILLS CHECK **2A EXTRA** is on the CD

2.3 Arithmetic series

Arithmetic series, including the formula for the sum of the first n natural numbers.

The sequence defined by $u_n = 4n + 1$ is 5, 9, 13, 17, 21, ...

This is an example of an **arithmetic sequence**, where each successive term is obtained from the previous one by *adding* a constant amount, called the **common difference**.

Adding the terms of an arithmetic sequence gives an **arithmetic series**, for example $5 + 9 + 13 + 17 + 21 + \cdots$

Note:
The iterative formula for this sequence is $x_{n+1} = x_n + 4$, with $x_1 = 5$.

General expression for u_n

In an arithmetic series, the first term, u_1 is usually denoted by a and the common difference is denoted by d.

The terms are as follows:

$u_1 = a$

$u_2 = a + d$

$u_3 = (a + d) + d = a + 2d$

$u_4 = (a + 2d) + d = a + 3d$

Continuing the pattern, $u_n = a + (n - 1)d$.

In the series $5 + 9 + 13 + 17 + 21 + \cdots$, $a = 5$ and $d = 4$,

so $u_n = 5 + (n - 1) \times 4$

$= 5 + 4n - 4$

$= 4n + 1$, as expected.

Example 2.7 The third term in an arithmetic series is 8 and the eighth term is 18. Find

a the first term and the common difference,

b the 23rd term,

c the nth term.

Step 1: Use the nth term to form two equations in a and d.

a

$a + 2d = 8$ ①

$a + 7d = 18$ ②

② − ①

$5d = 10$

Step 2: Solve the equations.

$d = 2$

Substituting in ①: $a + 4 = 8$

$a = 4$

The first term is 4 and the common difference is 2.

Step 3: Use the nth term formula.

b $u_{23} = a + 22d = 4 + 22 \times 2 = 48$

c nth term $= a + (n - 1)d$

$= 4 + (n - 1) \times 2$

$= 4 + 2n - 2$

$= 2n + 2$

Sum of first n terms, S_n

The sum of the first n terms of an arithmetic series, with first term a and common difference d, is written S_n and is given by

$$S_n = \frac{n}{2}(2a + (n - 1)d)$$

Alternatively, if you know the first term and the last term, l, use

$$S_n = \frac{n}{2}(a + l)$$

Example 2.8 Find the sum of the first 20 terms of the arithmetic series
$$4 + 7 + 10 + 13 + \cdots$$

Step 1: Define a, d and n. $\quad a = 4, \; d = 3, \; n = 20$

Step 2: Use the formula for S_n.

$$S_n = \frac{n}{2}(2a + (n-1)d)$$

$$S_{20} = \frac{20}{2}(2 \times 4 + 19 \times 3) = 650$$

Example 2.9 The first term of an arithmetic series is 6 and the eighth term is twice the third term. Find the sum of the first ten terms.

Step 1: Use the nth term formula to write u_8 and u_3 in terms of d.

$$a = 6, \; u_8 = 6 + 7d, \; u_3 = 6 + 2d$$

Step 2: Use the given relationship to find d.

$$u_8 = 2u_3 \Rightarrow 6 + 7d = 2(6 + 2d)$$
$$6 + 7d = 12 + 4d$$
$$d = 2$$

Step 3: Use the formula for S_n.

$$S_{10} = \frac{10}{2}(2 \times 6 + 9 \times 2) = 150$$

Example 2.10 Ben's aunt gave him money on his birthday every year from his 15th birthday to his 30th birthday.

She gave him £200 on his 15th birthday. How much did she give him in total if on each subsequent year she gave him

a £100,

b £100 more than on his previous birthday.

Step 1: Identify the type of sequence.

a Ben's age: 15   16   17   18   ...   30

Total given: £200   £300   £400   £500   ...   ?

This is an **arithmetic sequence**, $a = 200$, $d = 100$, $n = 16$.

> **Tip:**
> In **a**, each term is the total amount Ben had been given on and before the particular birthday.

> **Tip:**
> Take care when working out n.

Step 2: Identify the term required and use an appropriate formula.

The total amount Ben had received by his 30th birthday is given by the 16th term of the sequence.

$$u_n = a + (n-1)d$$
$$u_{16} = 200 + 15 \times 100 = 1700$$

Ben had been given £1700.

Step 1: Identify the type of series.

b Ben's age: 15   16   17   18   ...   30

Amount given each year: £200   £300   £400   £500   ...   ?

So, by his 30th birthday, the total amount given to Ben is
$200 + 300 + 400 + \cdots +$ the amount given on his 30th birthday

> **Tip:**
> In **b**, the total amount is found by adding the amounts given each year.

This is an **arithmetic series**, $a = 200$, $d = 100$, $n = 16$.

Step 2: Identify the sum required and use an appropriate formula.

The total amount is the sum of the series.

$$S_{16} = \frac{16}{2}(2 \times 200 + 15 \times 100) = 15\,200$$

Ben had been given £15 200.

Using Σ notation to describe an arithmetic series

The nth term in an arithmetic sequence can be written in the form $dn + c$, where d is the common difference and c is a constant. Hence an arithmetic series, with common difference d, can be written

$$\sum_{r=1}^{n} (dr + c).$$

For example $\sum_{r=1}^{5} (4r + 1) = 5 + 9 + 13 + 17 + 21 = 65.$

Recall:
Σ means 'the sum of' (see Section 2.2).

Note:
Σ notation can be used for other series, such as geometric series. See Section 2.4.

Note:
The series is arithmetic, with common difference 4.

Example 2.11 Evaluate $\sum_{r=1}^{14} (3r - 2)$.

Step 1: Substitute integer values of r, from the lower to higher number.

$$\sum_{r=1}^{14} (3r - 2) = 1 + 4 + 7 + \cdots + 40$$

Step 2: Identify the series and use an appropriate formula.

This is an arithmetic series with $a = 1$, $d = 3$, $l = 40$, $n = 14$.

$$\sum_{r=1}^{14} (3r - 2) = S_{14} = \frac{14}{2}(1 + 40) = 287$$

Tip:
Work out the first few terms and the last term.

Tip:
Use $S_n = \dfrac{n}{2}(a + l)$.

Example 2.12 Show that $1 + 2 + 3 + \cdots + n = \dfrac{n(n + 1)}{2}$.

Step 1: Identify the type of series.

This is an arithmetic series, with n terms, where $a = 1$, $d = 1$, $l = n$.

Step 2: Find S_n.

$$S_n = \frac{n}{2}(a + l) = \frac{n}{2}(1 + n) = \frac{n(n + 1)}{2}$$

Note:
You could use
$S_n = \dfrac{n}{2}(2a + (n - 1)d)$.

The result in Example 2.12 can be expresssed as follows:

The **sum of the first n natural numbers** is given by:

$$\sum_{r=1}^{n} r = \frac{n(n + 1)}{2}$$

Note:
Natural numbers are the counting numbers 1, 2, 3, 4, ...

Tip:
This formula will be given in the examination.

Example 2.13 Evaluate **a** $\sum_{r=1}^{100} r$ **b** $\sum_{r=20}^{50} r$.

Step 1: Use the Σr formula. **a** $\sum_{r=1}^{100} r = 1 + 2 + 3 + \cdots + 100 = \dfrac{100 \times 101}{2} = 5050$

Step 2: Use the Σr formula in two stages, subtracting unwanted terms. **b** $\sum_{r=20}^{50} r = \sum_{r=1}^{50} r - \sum_{r=1}^{19} r$

$$= \frac{50 \times 51}{2} - \frac{19 \times 20}{2}$$

$$= 1275 - 190$$

$$= 1085$$

Tip:
The series is
$20 + 21 + \ldots + 50$.

Tip:
To use the Σr formula, the lowest value of r must be 1.

The following relationship can be used to sum any arithmetic series:

$$\sum_{r=1}^{n} (dr + c) = d\sum_{r=1}^{n} r + nc$$

Example 2.14 Evaluate $\displaystyle\sum_{r=1}^{24} (3r + 2)$.

Method 1

Step 1: Expand, using Σ notation.

$$\sum_{r=1}^{24} (3r + 2) = 3\sum_{r=1}^{24} r + 24 \times 2$$

Tip:
$n = 24$.

Step 2: Apply the formula for Σr.

$$= 3 \times \frac{24 \times 25}{2} + 48$$

$$= 948$$

Method 2

Step 1: Write out the first few and the last terms.

Alternatively, you can use the expanded series and the formula for the sum of an arithmetic series:

Tip:
Choose the method you prefer.

$$\sum_{r=1}^{24} (3r + 2) = 5 + 8 + 11 + \cdots + 74$$

Tip:
Substitute $r = 1, 2, 3, \ldots, 24$.

Step 2: Identify the series.

This is an arithmetic series with $a = 5, d = 3, l = 74, n = 24$.

Step 3: Use the appropriate formula.

$$S_n = \frac{n}{2}(a + l)$$

$$= \frac{24}{2}(5 + 74)$$

$$= 948$$

SKILLS CHECK **2B: Arithmetic series**

1 Find the common difference and the sum of the first 20 terms of the following series:

   a $12 + 17 + 22 + \cdots$ **b** $-2 - 5 - 8 - \cdots$

2 Find the nth term and the sum of the first 200 terms of the arithmetic series $\frac{1}{2} + \frac{3}{2} + \frac{5}{2} + \frac{7}{2} + \cdots$

3 The first term of an arithmetic series is 8 and the seventh term is 26. Find

   a the common difference,   **b** the nth term,   **c** the sum of the first 25 terms.

4 The third, fourth and fifth terms of an arithmetic series are $(4 + x)$, $2x$ and $(8 - x)$ respectively.

   a Find the value of x.

   b Find the first term and the common difference.

   c Find the sum of the first 30 terms of the series.

5 The cost to a company of training a student is £1000 for the first student. The cost is then reduced by £50 for the second student, by a further £50 for the third student and so on, so that the cost for the second student is £950, for the third student is £900 and so on.

   a How much will it cost to train the 20th student?

   b How much will it cost the company to train 20 students?

6 Evaluate   **a** $\displaystyle\sum_{r=1}^{38} r$ **b** $\displaystyle\sum_{r=1}^{18} (7r - 1)$.

7 Evaluate $\displaystyle\sum_{r=12}^{20} (\tfrac{1}{2}r + 3)$.

8 Find n if $\displaystyle\sum_{r=1}^{2n}(4r-1) = \sum_{r=1}^{n}(3r+59)$.

9 The nth term of an arithmetic sequence is u_n, where $u_n = 6 + 2n$.

  a Find the values of u_1, u_2 and u_3.

  b Write down the common difference of the arithmetic sequence.

  c Find the value of n for which $u_n = 46$.

  d Evaluate $\displaystyle\sum_{n=1}^{20} u_n$.

10 Sharon borrowed £5625 on an interest-free loan. She paid back £25 at the end of the first month, then increased her payment by £50 in each subsequent month, paying £75 at the end of the second month, £125 at the end of the third month and so on.

  a How many months did she take to pay off the loan?

  b What was the amount of her final month's repayment?

SKILLS CHECK **2B EXTRA** is on the CD

2.4 Geometric series

The sum of a finite geometric series; the sum to infinity of a convergent $(-1 < r < 1)$ geometric series.

The sequence defined by $u_n = 3(2^{n-1})$ is 3, 6, 12, 24, 48, …

This is an example of a **geometric sequence**, where each term is obtained from the previous one by *multiplying* by a constant. The constant is called the **common ratio**.

Adding the terms of a geometric sequence gives a **geometric series**, for example $3 + 6 + 12 + 24 + 48 + \cdots$

Note:
The recurrence relation for this geometric sequence is $x_{n+1} = 2x_n$ with $x_1 = 3$.

Tip:
To find the common ratio, divide any term by the previous term.

General expression for u_n

In a geometric series the first term u_1 is usually denoted by a and the common ratio is denoted by r.

The terms are as follows:

$u_1 = a$

$u_2 = ar$

$u_3 = ar^2$

$u_4 = ar^3$

Continuing the pattern, $u_n = ar^{n-1}$.

Tip:
The formula for u_n will be given in the examination but it is useful to learn it.

Example 2.15 Find the eighth term of the geometric series $3 + 6 + 12 + \cdots$

Step 1: Define a and r. $a = 3$, $r = 2$

Step 2: Use the formula for u_n. $u_n = 3(2^{n-1})$

$\Rightarrow u_8 = 3(2^7) = 384$

Example 2.16 A geometric series has second term 12 and fourth term 48. Find the seventh term, given that the first term is negative.

Step 1: Use the nth term formula to form two equations in a and r.

$u_2 = 12 \Rightarrow ar = 12$ ①
$u_4 = 48 \Rightarrow ar^3 = 48$ ②

Step 2: Solve the equations.

② ÷ ①: $\dfrac{ar^3}{ar} = \dfrac{48}{12}$

$r^2 = 4$

$r = \pm 2$

Substituting in ①:

When $r = 2$, $2a = 12 \Rightarrow a = 6$.

(This is not applicable, since $a < 0$.)

When $r = -2$, $-2a = 12 \Rightarrow a = -6$.

> **Tip:**
> Test which value of r gives a negative value of a.

Step 3: Use u_n formula with $n = 7$.

Since $a < 0$, $r = -2$ and $a = -6$.

$u_7 = ar^6 = (-6) \times (-2)^6 = -384$

Example 2.17 In January 2001, an investor put £1000 into a savings account with a fixed interest rate of 4% per annum. Interest is added to the account on 31 December each year and no further capital is invested.

  a By what factor is the amount in the account increased when interest is added?

  b How much will be in the account, to the nearest £, when interest has been added on 31 December 2015?

Step 1: Calculate the multiplying factor.

  a For an interest rate of 4%, the amount grows by $(1 + \frac{4}{100})$ each year, that is, by a factor of 1.04.

Step 2: Apply the multiplying factor for the appropriate number of years.

  b Amount in account:

January 2001	1000
December 2001	1000×1.04
December 2002	$1000 \times 1.04 \times 1.04 = 1000(1.04)^2$
$\vdots$	$\vdots$ $\qquad$ $\vdots$
December 2015	$1000(1.04)^{15}$ $\quad = 1800.94\ldots$

> **Note:**
> The amount in December 2015 is the 16th term in a geometric sequence, first term 1000, common ratio 1.04.

The amount in the account on 31 December 2015 is £1801 (nearest £).

Sum of the first n terms, S_n

The sum of the first n terms of a geometric series is:

$$S_n = \frac{a(1 - r^n)}{1 - r}$$

> **Note:**
> This formula will be given in the examination.

Example 2.18 Calculate the sum of the first ten terms of the geometric series
$2 + 6 + 18 + 54 + \cdots$

Step 1: Define a, r and n.

$a = 2$, $r = 3$, $n = 10$

Step 2: Use the formula for the sum S_n.

$S_{10} = \dfrac{2(1 - 3^{10})}{1 - 3}$

$= 59\,048$

> **Tip:**
> You could use $S_n = \dfrac{a(r^n - 1)}{r - 1}$.

Sum to infinity of a convergent geometric series

If r lies between -1 and 1, that is, $|r| < 1$, then as n gets larger, the terms in a geometric series get smaller. The sum of the series tends to a limiting value.

Note:
When $|r| < 1$, as $n \to \infty$, $ar^{n-1} \to 0$.

This is known as the **sum to infinity**, S_∞, where $S_\infty = \dfrac{a}{1-r}$.

Example 2.19 The third term of a geometric series is $\frac{8}{3}$ and the sixth term is $\frac{64}{81}$.

Find **a** the sum of the first twenty terms, **b** the sum to infinity.

Step 1: Use the nth term formula to write u_3 and u_6 in terms of a and r.

$$u_3 = \tfrac{8}{3} \Rightarrow ar^2 = \tfrac{8}{3} \qquad \textcircled{1}$$

$$u_6 = \tfrac{64}{81} \Rightarrow ar^5 = \tfrac{64}{81} \qquad \textcircled{2}$$

Step 2: Solve to find a and r.

$$\textcircled{2} \div \textcircled{1} \qquad \frac{ar^5}{ar^2} = \frac{\frac{64}{81}}{\frac{8}{3}}$$

Tip:
Work in fractions.

$$\Rightarrow \qquad r^3 = \tfrac{8}{27}$$

$$r = \sqrt[3]{\tfrac{8}{27}} = \tfrac{2}{3}$$

Substituting in $\textcircled{1}$:

$$a\left(\tfrac{2}{3}\right)^2 = \tfrac{8}{3}$$

$$a = \tfrac{8}{3} \div \left(\tfrac{2}{3}\right)^2 = 6$$

Step 3: Use the formula for S_n.

$$S_{20} = \frac{a(1-r^{20})}{1-r} = \frac{6\left(1 - \frac{2}{3}^{20}\right)}{1 - \frac{2}{3}} = 17.9945\ldots = 18 \text{ (2 s.f.)}$$

Step 4: Use the sum-to-infinity formula.

Since $r = \tfrac{2}{3}$, $|r| < 1$ and the series converges.

$$S_\infty = \frac{a}{1-r} = \frac{6}{1 - \frac{2}{3}} = 18$$

Tip:
This is what you might expect, given the value for S_{20}.

Using Σ notation to describe a geometric series

A geometric series with common ratio r can be written in the form

$$\sum_{i=1}^{n} p(r^i),$$ where p and r are numbers.

Tip:
Look for a power of r in the general term.

For example, $\displaystyle\sum_{i=1}^{4} 5(2^i) = 10 + 20 + 40 + 80 = 150$.

This series has common ratio 2, as expected.

Note:
In this text, i has been used for the summation to avoid confusion with the common ratio r. Be careful: this may not always be the case!

Example 2.20 Evaluate $\displaystyle\sum_{i=1}^{9} 3(2^i)$.

Step 1: Expand the series and identify a, r and n.

$$\sum_{i=1}^{9} 3(2^i) = 3(2^1) + 3(2^2) + \cdots + 3(2^9)$$

This is a geometric series with $a = 3 \times 2 = 6$, $r = 2$ and $n = 9$, so S_9 is required.

Step 2: Apply geometric series formula for S_n.

Using $\qquad S_n = \dfrac{a(1-r^n)}{1-r}$

$$\sum_{i=1}^{9} 3(2^i) = S_9 = \frac{6(1-2^9)}{1-2} = 3066$$

Example 2.21 Evaluate $\sum_{i=0}^{\infty} \left(\frac{1}{2}\right)^i$.

Tip:
The sum to infinity is required.

Step 1: Expand the series and identify a, r and n.

$$\sum_{i=0}^{\infty} \left(\tfrac{1}{2}\right)^i = \left(\tfrac{1}{2}\right)^0 + \left(\tfrac{1}{2}\right)^1 + \left(\tfrac{1}{2}\right)^2 + \cdots = 1 + \tfrac{1}{2} + \tfrac{1}{4} + \cdots$$

This is a geometric series with $a = 1$, $r = \frac{1}{2}$. S_∞ is required.

Step 2: Apply geometric series formula for S_∞.

$$\sum_{i=0}^{\infty} \left(\tfrac{1}{2}\right)^i = \frac{a}{1-r} = \frac{1}{1-\frac{1}{2}} = 2$$

Tip:
Check that $|r| < 1$ is satisfied.

SKILLS CHECK 2C: Geometric series

1 For the following geometric series, find
  i the seventhth term   **ii** the sum of the first seven terms   **iii** the sum to infinity (if possible):

  a $2 + 10 + 50 + \cdots$ **b** $7 + \frac{7}{2} + \frac{7}{4} + \cdots$ **c** $1 - 2 + 4 - 8 + \cdots$

2 Find the sum of the first ten terms of the geometric series $-2 - 6 - 18 - \cdots$

3 A property was valued at £70 000 at the start of 2001. If the projected increase in value is 3% per year, find the projected value of the property at the start of 2020. Give your answer to the nearest £1000.

4 A geometric series has first term a and common ratio r, where $r > 0$. The third term is $\frac{5}{2}$ and the seventh term is $\frac{5}{512}$.

  a Find the values of a and r.   **b** Find the sum to infinity of the series.

5 The sum to infinity of a geometric series is $\frac{3}{4}$ and the sum of the first two terms is $\frac{2}{3}$. The common ratio of the series is negative.

  a Find the common ratio.

  b Find the *exact* difference between the sum of the first five terms and the sum to infinity.

6 Evaluate   **a** $\sum_{i=1}^{9} 2(4^i)$ **b** $\sum_{i=1}^{\infty} \left(\frac{3}{4}\right)^i$.

7 Evaluate $\sum_{i=1}^{6} (7^i + 1)$.

8 On her first birthday, Belinda is given £1. In each subsequent year, she is given double the amount that she received in the previous year, so that she receives £2 on her second birthday, £4 on her third birthday and so on.

  a How much does she receive on her tenth birthday?

  b How much, in total, has she received when she is 10?

9 A geometric series has first term 2 and second term $2\sqrt{2}$.

  a Find the seventh term.

  b The sum of the first five terms is $p + q\sqrt{2}$. Find the values of p and q.

10 Evaluate $\sum_{i=1}^{\infty} 2^{-i}$.

SKILLS CHECK 2C EXTRA is on the CD

The binomial expansion of $(1 + x)^n$ for positive integer n.

Notice the pattern in these **binomial expansions**:

$$(a + b)^0 = \qquad\qquad 1$$
$$(a + b)^1 = \qquad\qquad a + b$$
$$(a + b)^2 = \qquad\qquad a^2 + 2ab + b^2$$
$$(a + b)^3 = \qquad\qquad a^3 + 3a^2b + 3ab^2 + b^3$$
$$(a + b)^4 = \qquad\qquad a^4 + 4a^3b + 6a^2b^2 + 4ab^3 + b^4$$
$$(a + b)^5 = \qquad a^5 + 5a^4b + 10a^3b^2 + 10a^2b^3 + 5ab^4 + b^5$$

Note:
A binomial expansion contains two terms in the bracket.

Note:
The powers of a descend and the powers of b ascend. In any term, the powers of a and b add up to the power of the bracket.

The coefficients of the terms can be found using **Pascal's triangle.** The first few lines are as follows:

```
              1
           1     1
        1     2     1
     1     3     3     1
  1     4     6     4     1
1     5    10    10     5     1
```

Tip:
$a = 1a^1$

Tip:
A number is formed by adding the two numbers immediately above it.

6 + 4
10

Example 2.22 Expand $(3 + 2x)^4$ in ascending powers of x.

Step 1: Recall the appropriate line of Pascal's triangle.

The appropriate line of Pascal's triangle is $\quad 1 \quad 4 \quad 6 \quad 4 \quad 1.$

Step 2: Compare with $(a + b)^n$ and expand.

$$(3 + 2x)^4 = 3^4 + 4(3^3)(2x) + 6(3^2)(2x)^2 + 4(3)(2x)^3 + (2x)^4$$
$$= 81 + 216x + 216x^2 + 96x^3 + 16x^4$$

Using $\binom{n}{r}$ to find the coefficients

For positive integer values of n, $\binom{n}{r} = {}^nC_r = \dfrac{n!}{(n-r)!r!}$. It can be calculated using factorial notation, for example

$$\binom{5}{2} = \frac{5!}{2!3!} = \frac{5\times4\times3\times2\times1}{2\times1\times3\times2\times1} = 10.$$

Note:
$n! = n(n-1)(n-2)$
$\qquad\qquad \dots \times 3 \times 2 \times 1,$
$5! = 5 \times 4 \times 3 \times 2 \times 1$
$\quad = 120$

Alternatively use the $\boxed{nCr}$ button on your calculator and key in $\boxed{5}$ $\boxed{nCr}$ $\boxed{2}$ $\boxed{=}$.

Tip:
Make sure that you know how to find $\boxed{nCr}$ on *your* calculator.

These expanded results are very useful:

$$\binom{n}{0} = 1 \qquad \binom{n}{1} = n \qquad \binom{n}{2} = \frac{n(n-1)}{2!}$$

$$\binom{n}{3} = \frac{n(n-1)(n-2)}{3!} \qquad \dots \qquad \binom{n}{n} = 1$$

For example, $\binom{10}{3} = \dfrac{10\times9\times8}{3!} = \dfrac{720}{6} = 120.$

Binomial expansion formula for $(a + b)^n$

In general, for positive integer values of n,

$$(a + b)^n = a^n + \binom{n}{1} a^{n-1}b + \binom{n}{2} a^{n-2}b^2 + \cdots + \binom{n}{r} a^{n-r}b^r + \cdots + b^n$$

$$= a^n + na^{n-1}b + \frac{n(n-1)}{2!} a^{n-2}b^2 + \cdots + b^n$$

In the examination, this formula will be given in the booklet.

Example 2.23 Expand $(2x + 3y)^4$.

Step 1: Compare with $(a + b)^n$ and expand using the formula.

Comparing with $(a + b)^n$, $a = 2x$, $b = 3y$, $n = 4$:

$$(2x + 3y)^4 = (2x)^4 + \binom{4}{1}(2x)^3(3y) + \binom{4}{2}(2x)^2(3y)^2 + \binom{4}{3}(2x)(3y)^3 + (3y)^4$$

Step 2: Simplify.

$$= (2x)^4 + 4(2x)^3(3y) + 6(2x)^2(3y)^2 + 4(2x)(3y)^3 + (3y)^4$$

$$= 16x^4 + 96x^3y + 216x^2y^2 + 216xy^3 + 81y^4$$

Tip:
To find the coefficients, use ^4C_r on your calculator, use the expanded form or, easiest of all, recall the appropriate line of Pascal's triangle.

Example 2.24 Given that $(2 - y)^9 = A + By + Cy^2 + \cdots$, find A, B and C.

Step 1: Compare with $(a + b)^n$ and expand using the formula.

Comparing with $(a + b)^n$, $a = 2$, $b = (-y)$, $n = 9$.

$$(2 - y)^9 = 2^9 + \binom{9}{1}2^8(-y) + \binom{9}{2}2^7(-y)^2 + \cdots$$

Step 2: Simplify.

$$= 2^9 + 9(2^8)(-y) + 36(2^7)(-y)^2 + \cdots$$

$$= 512 - 2304y + 4608y^2 + \cdots$$

$A = 512$, $B = -2304$ and $C = 4608$.

Tip:
You need to calculate the first three terms of the series.

Tip:
$\binom{9}{2} = \frac{9 \times 8}{2 \times 1} = 36$

Example 2.25 Find the coefficient of x^4 in the expansion of $(2 + \frac{1}{2}x)^5$.

Step 1: Identify the term required and calculate the coefficient.

The term required is $\binom{5}{4} 2^1(\frac{1}{2}x)^4 = 5 \times 2 \times \frac{1}{16}x^4 = \frac{5}{8}x^4$.

The coefficient of x^4 is $\frac{5}{8}$.

Tip:
Take care with $(\frac{1}{2}x)^4$. Remember to raise $\frac{1}{2}$ to the power 4 as well as x.

Binomial expansion formula for $(1 + x)^n$

If the first term in the bracket is 1, you can use a simplified version of the expansion formula.

For positive integer values of n:

$$(1 + x)^n = 1 + \binom{n}{1}x + \binom{n}{2}x^2 + \binom{n}{3}x^3 + \cdots + \binom{n}{r}x^r + \cdots + x^n$$

$$= 1 + nx + \frac{n(n-1)}{2!}x^2 + \frac{n(n-1)(n-2)}{3!}x^3 + \cdots + x^n$$

Example 2.26 **a** Expand $(1 - 2x)^8$ in ascending powers of x, as far as the term in x^3.

b By letting $x = 0.01$, use the first four terms to find an approximate value for 0.98^8.

Step 1: Compare with the general formula for $(1 + x)^n$, expand and simplify.

Step 2: Substitute the x-value and calculate.

a $(1 - 2x)^8 = 1 + 8(-2x) + \dfrac{8\times7}{2!}(-2x)^2 + \dfrac{8\times7\times6}{3!}(-2x)^3 + \cdots$

$= 1 - 16x + 112x^2 - 448x^3 + \cdots$

b When $x = 0.01$, $1 - 2x = 1 - 0.02 = 0.98$.

So $0.98^8 \approx 1 - 16(0.01) + 112(0.01)^2 - 448(0.01)^3$

$= 0.850752$

$= 0.8508$ (4 d.p.)

Tip:
Replace n with 8 and 'x' with '$-2x$' in the formula for $(1 + x)^n$.

Note:
On the calculator $0.98^8 = 0.85076\ldots$ so the approximation is correct to 4 d.p.

Example 2.27 In the expansion of $(1 + kx)^n$, where k and n are positive integers, the coefficient of x is 15 and the coefficient of x^2 is 90.

a Show that $k = \dfrac{15}{n}$ and find n and k.

b Hence find the term in x^3.

Step 1: Compare with the general formula for $(1 + x)^n$.
Step 2: Expand and simplify.

Step 3: Compare the terms and solve for the unknowns.

a $(1 + kx)^n = 1 + n(kx) + \dfrac{n(n - 1)}{2!}(kx)^2 + \cdots$

$= 1 + nkx + \dfrac{n(n - 1)}{2!}k^2x^2 + \cdots$

Coefficient of x: $\quad nk = 15 \Rightarrow k = \dfrac{15}{n}$ ①

Coefficient of x^2: $\quad \dfrac{n(n - 1)}{2!}k^2 = 90$ ②

Substituting for k from ①:

$\dfrac{n(n - 1)}{2}\left(\dfrac{15}{n}\right)^2 = 90$

$\dfrac{225n(n - 1)}{2n^2} = 90$

$225(n - 1) = 180n$

$225n - 225 = 180n$

$45n = 225$

$n = 5$

Substituting in ①: $k = \dfrac{15}{5} = 3$, so $n = 5$, $k = 3$.

Recall:
Simultaneous equations (Core 1 Section 1.7).

Tip:
Cancel n, then multiply both sides by $2n$.

b Term in x^3: $\quad \dbinom{5}{3}(3x)^3 = 10 \times 3^3 \times x^3 = 270x^3$

SKILLS CHECK **2D: Binomial expansions**

1 The polynomial p(x) is given by $(2 - 3x)^4$. Find the binomial expansion of p(x), simplifying your terms.

2 Simplifying your terms, find the first four terms in the expansion of $(1 + 4y)^7$, in ascending powers of y.

3 **a** Expand $(3 - 2x)^5$ in ascending powers of x up to the term in x^2.

   b Find the values of A, B and C, where $(5 + 2x)(3 - 2x)^5 = A + Bx + Cx^2 + \cdots$

4 a Expand $(1 + 2x)^6$ in ascending powers of x up to the term in x^3.

   b Using your expansion, find an approximation for $(1.02)^6$, correct to four decimal places. You must write down sufficient working to show how you obtained your answer.

5 Using the first four terms, in ascending powers of x, of the expansion of $(1 - 4x)^7$, find an approximate value for $(0.996)^7$, to a suitable degree of approximation. You must write down sufficient working to show how you obtained your answer.

6 Expand and simplify $(\sqrt{2} + \sqrt{3})^4 - (\sqrt{2} - \sqrt{3})^4$, leaving your answer in the form $a\sqrt{6}$, where a is a positive integer.

7 The coefficient of x^2 is $\frac{3}{8}$ in the expansion of $\left(1 + \dfrac{x}{n}\right)^n$. Find the value of n.

8 It is given that $(1 + kx)^n = 1 - 4x + 7x^2 + \cdots$

   a Find n and k.

   b Hence find the term in x^3.

9 a Expand $(1 + ax)^6$ in ascending powers of x up to and including the term in x^2.

   b In the expansion $(1 + bx)(1 + ax)^6$, the coefficients of x and x^2 are 20 and 171 respectively. Find a and b, given that they are integers.

10 a Write down the first four terms, in ascending powers of x, in the expansion of $(1 - 3x)^5$.

   b Find the coefficient of x^3 in the expansion of $(1 + x)(1 - 3x)^5$.

SKILLS CHECK **2D EXTRA** is on the CD

Examination practice   Sequences and series

1 An arithmetic series has sixth term 28 and tenth term 44.

   a Find the first term and the common difference.

   b Find the sum of the first 50 terms of the series.   [AQA (B) Nov 2003]

2 An arithmetic series has first term a and common difference d. The sum of the first 19 terms is 266.

   a Show that $a + 9d = 14$.

   b The sum of the fifth and eighth terms is 7. Find the values of a and d.   [AQA (B) Jan 2003]

3 The first term of an arithmetic series is 7. The tenth term is 43.

   a Find the common difference.

   b Find the sum of the first fifty terms of the series.

   c The kth term has a value greater than 1000.

      i Show that $4k > 997$.

      ii Find the least possible value of k.   [AQA (B) Jan 2004]

4 a Find the sum of the 16 terms of the arithmetic series

$$2 + 5 + 8 + \ldots + 47.$$

b An arithmetic sequence $u_1, u_2, u_3, \ldots$ has rth term u_r, where

$$u_r = 50 - 3r.$$

    i Write down the values of u_1, u_2, u_3 and u_4.

    ii Show that the sequence has exactly 16 positive terms.   [AQA (A) June 2002]

5 The nth term of an arithmetic sequence is u_n, where

$$u_n = 10 + 0.5n.$$

a Find the values of u_1 and u_2.

b Write down the common difference of the arithmetic sequence.

c Find the value of n for which $u_n = 25$.

d Evaluate $\displaystyle\sum_{n=1}^{30} u_n$.   [AQA (A) Nov 2002]

6 A pipeline is to be constructed under a lake. It is calculated that the first mile will take 15 days to construct. Each further mile will take 3 days longer than the one before, so the 1st, 2nd and 3rd miles will take 15, 18 and 21 days, respectively, and so on.

a Find the nth term of the arithmetic sequence 15, 18, 21, ...

b Show that the total time taken to construct the first n miles of the pipeline is $\frac{3}{2}n(n + 9)$ days.

c Calculate the total length of pipeline that can be constructed in 600 days.   [AQA (A) Jan 2002]

7 The second term of a geometric series is 24 and the fifth term is 3.

a Show that the common ratio of the series is $\frac{1}{2}$.

b Find the first term of the series.

c Find the sum to infinity of the series.   [AQA (B) Jan 2002]

8 The first four terms of a geometric sequence are

$$10, 9, 8.1, 7.29.$$

a Show that the common ratio of the sequence is 0.9.

b Find the nth term.

c Show that the sum of the first 25 terms is approximately 92.8.

d Find the sum to infinity.   [AQA (A) Jan 2003]

9 The sum to infinity of a convergent geometric series is $5a$, where a is the first term of the series.

a Show that the common ratio of the series is $\frac{4}{5}$.

b The third term of the series is 64.

    i Find the first term of the series.

    ii Find the **exact** decimal value for the sum of the first six terms.   [AQA (B) Nov 2003]

10 A geometric series has first term a and common ratio r, with $a > 0$ and $r > 0$.

a The second term of the series is 4 and the eighth term is $\frac{1}{2}$. Show that $r = \dfrac{1}{\sqrt{2}}$ and find the value of a in surd form.

b Find the sum to infinity of the series in the form $k(\sqrt{2} + 1)$, and state the value of the integer k.

[AQA (B) Jan 2001]

11 a The first three terms of a geometric sequence are a, b, c. Each term represents an increase of p per cent on the preceding term.

i Show that the common ratio is $\left(1 + \dfrac{p}{100}\right)$.

ii It is given that $a = 2000$. Express b and c in terms of p.

b A deposit of £2000 is put into a bank account. After each year, the balance in the account is increased by p per cent. There are no other deposits or withdrawals. After two years the balance is £2332.80.

i Show that $p = 8$.

ii Given that after n years the balance is £u_n, write down an expression for u_n in terms of n.

iii Use your answer to part **b ii** to find the balance after 10 years. [AQA (A) June 2003]

12 a A geometric series has first term 1200 and common ratio r. Write down the second and third terms of the series in terms of r.

b A total of £11 700 is to be shared amongst three people. The values of the three shares are the first three terms in a geometric series with common ratio r. The smallest share is to be £1200.

i Show that r satisfies the equation
$$4r^2 + 4r - 35 = 0.$$

ii Hence, or otherwise, find the value of the largest share. [AQA (B) June 2002]

13 a Find the sum of the three hundred integers from 101 to 400 inclusive.

b Find the sum of the geometric series $2 + 6 + 18 + \cdots + 2 \times 3^{n-1}$, giving your answer in the form $p^n - q$, where p and q are integers. [AQA (A) June 2001]

14 The polynomial $p(x)$ is given by $p(x) = (x + 2)^5$. Find the binomial expansion of $p(x)$, simplifying your terms. [AQA (B) June 2003]

15 a Write down the first four terms in ascending powers of x in the expansion of
$$(1 + x)^8,$$
simplifying your coefficients as much as possible.

b Find the coefficient of x^3 in the expansion of $(3 - 2x)(1 + x)^8$. [AQA (B) Jan 2002]

16 A polynomial can be expressed in the form
$$p(x) = (x - 2)^4 - (x + 1)^3.$$
Use binomial expansions to express $p(x)$ in the form $x^4 + ax^3 + bx^2 + cx + d$, where a, b, c and d are integers. [AQA (B) June 2002]

3 Trigonometry

3.1 Sine and cosine rules

The sine and cosine rules.

The sine and cosine rules are used to find lengths and angles in a triangle.

To apply them, you need to label your triangle as follows:

Label the vertices with upper case letters, for example A, B and C. Then label the side opposite each vertex with the corresponding lower case letter, so side a is opposite angle A, side b is opposite angle B and side c is opposite angle C.

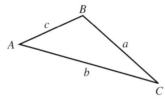

Note:
The triangle can be any size and shape.

Note:
A **vertex** is where two lines meet. The plural is **vertices**.

The sine rule

To find a **length**, use

$$\frac{a}{\sin A} = \frac{b}{\sin B} = \frac{c}{\sin C} \qquad \textit{Format (1)}$$

To find an **angle**, use

$$\frac{\sin A}{a} = \frac{\sin B}{b} = \frac{\sin C}{c} \qquad \textit{Format (2)}$$

Tip:
You can use the sine rule to find:
• a side when you know the angles in the triangle and a side
• an angle when you know two sides and the angle opposite one of them.

Tip:
To proceed, you must know one complete ratio.

Note:
You must learn the sine rule.

Example 3.1 In triangle ABC, AC is 3.2 cm, angle ABC is 35° and angle BCA is 82°. Find AB, giving your answer to the nearest mm.

Step 1: Draw a carefully labelled sketch and include all known measures.

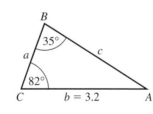

Step 2: Write down the sine rule in format (1) for finding a length.

Using the sine rule:

$$\frac{a}{\sin A} = \frac{b}{\sin B} = \frac{c}{\sin C}$$

Step 3: Substitute known values.

$$\frac{a}{\sin A} = \frac{3.2}{\sin 35°} = \frac{c}{\sin 82°}$$

Step 4: Choose the two relevant ratios and solve the equation.

$$\frac{c}{\sin 82°} = \frac{3.2}{\sin 35°}$$

$$c = \frac{3.2 \times \sin 82°}{\sin 35°}$$

$$= 5.5247...$$

So $AB = 5.5$ cm (to nearest mm)

Note:
You know length b and angle B. You are not asked anything about a and A so ignore the ratio involving them.

Note:
To give the answer to the nearest mm, you need to correct your value in cm to one decimal place.

Example 3.2   Find θ, giving your answer to the nearest degree.

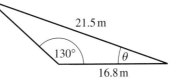

Step 1: Label the sketch carefully.

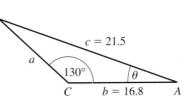

Step 2: Write down the sine rule in format (2) for finding an angle.

$$\frac{\sin A}{a} = \frac{\sin B}{b} = \frac{\sin C}{c}$$

Step 3: Substitute known values.

$$\frac{\sin \theta}{a} = \frac{\sin B}{16.8} = \frac{\sin 130°}{21.5}$$

Step 4: Choose the two relevant ratios and solve the equation.

$$\frac{\sin B}{16.8} = \frac{\sin 130°}{21.5}$$

$$\sin B = \frac{16.8 \times \sin 130°}{21.5}$$

$$= 0.59858...$$

$$B = 36.76...°$$

$$\Rightarrow \quad \theta = 180° - (130° + 36.76...°)$$

$$= 13.23...°$$

$$= 13° \text{ (to nearest degree)}$$

The cosine rule

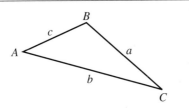

To find a **length**, for example a, use this formula for a^2:

$$a^2 = b^2 + c^2 - 2bc \cos A \qquad \qquad \textit{Format (1)}$$

Then square root to calculate a.

To find b, use

$$b^2 = a^2 + c^2 - 2ac \cos B$$

To find c, use

$$c^2 = a^2 + b^2 - 2ab \cos C$$

To find an **angle**, re-arrange the formulae as follows:

$$\cos A = \frac{b^2 + c^2 - a^2}{2bc} \qquad \qquad \textit{Format (2)}$$

$$\cos B = \frac{a^2 + c^2 - b^2}{2ac}$$

$$\cos C = \frac{a^2 + b^2 - c^2}{2ab}$$

Example 3.3 In triangle ABC, $AC = 6.2$ cm, $BC = 8.3$ cm and angle $ACB = 42°$.
Calculate AB, giving your answer correct to three significant figures.

Step 1: Draw a carefully
labelled sketch and include
all known measures.

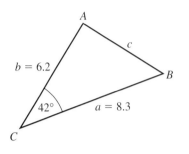

Tip:
The cosine rule is appropriate
as you know two sides and the
angle between them.

Step 2: Use the cosine rule
to find the missing length.

By the cosine rule

$$c^2 = a^2 + b^2 - 2ab \cos C$$

$$= 8.3^2 + 6.2^2 - 2 \times 8.3 \times 6.2 \times \cos 42°$$

$$= 30.84\ldots$$

$$c = \sqrt{30.84\ldots} = 5.553\ldots$$

$$AB = 5.55 \text{ cm (3 s.f.)}$$

Tip:
Find c^2 in one stage on your
calculator; do not press $\boxed{=}$
until you have entered cos 42.

Example 3.4 A triangle has sides of length 6 cm, 8 cm and 12 cm. Find the size
of the largest angle in the triangle, giving your answer to the
nearest 0.1°.

Step 1: Draw a carefully
labelled sketch and include
all known measures.

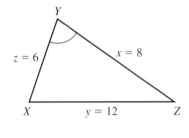

Tip:
The largest angle is opposite
the longest side.

Step 2: Use the cosine rule
to find the missing angle.

$$\cos Y = \frac{x^2 + z^2 - y^2}{2xz}$$

$$= \frac{8^2 + 6^2 - 12^2}{2 \times 8 \times 6}$$

$$= -0.458\ldots$$

$$Y = 117.3° \text{ (nearest 0.1°)}$$

Tip:
Giving your answer to the
nearest 0.1° is the same as
approximating to one decimal
place.

Calculator note:

There are several ways of entering the calculation. Make sure that the
method you use is a correct one.

Here is an example:

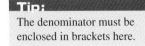

Tip:
The denominator must be
enclosed in brackets here.

3.2 Area of a triangle

The area of a triangle in the form $\frac{1}{2}ab\sin C$.

When you know **two sides** and the **angle between them**, you can use this formula to calculate the **area** of a triangle.

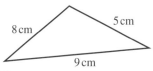

$$\text{Area} = \tfrac{1}{2}ab\sin C$$

You need to rearrange the formula if you know angle A or angle B as follows:

$$\text{Area} = \tfrac{1}{2}bc\sin A$$
$$\text{Area} = \tfrac{1}{2}ac\sin B$$

Example 3.5 In triangle PQR, $PQ = 4.2$ cm, $QR = 6.3$ cm and angle $PQR = 130°$. Calculate the area of the triangle, giving your answer to two significant figures.

Step 1: Draw a carefully labelled sketch and include all known measures.

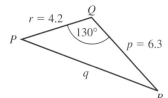

Step 2: Use the area formula.

$$\text{Area} = \tfrac{1}{2}\,pr\sin Q$$
$$= \tfrac{1}{2} \times 6.3 \times 4.2 \times \sin 130°$$
$$= 10.13\ldots$$
$$= 10\ \text{cm}^2\ (2\ \text{s.f.})$$

Example 3.6 The diagram shows a sketch of a triangle with sides of length 5 cm, 8 cm and 9 cm.

Calculate the area of the triangle, giving your answer correct to two significant figures.

Step 1: Draw a carefully labelled sketch and include all known measures.

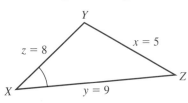

Step 2: Use the cosine rule to find an angle. Calculate an angle, using the cosine rule.

$$\cos X = \frac{y^2 + z^2 - x^2}{2yz}$$
$$= \frac{9^2 + 8^2 - 5^2}{2\times9\times8}$$
$$= 0.833\ldots$$
$$X = 33.557\ldots°$$

Step 3: Use the area formula.

$$\text{Area} = \tfrac{1}{2}\,yz\sin X$$
$$= \tfrac{1}{2} \times 9 \times 8 \times \sin 33.557..°$$
$$= 19.89\ldots$$
$$= 20\ \text{cm}^2\ (2\ \text{s.f.})$$

Tip:
It does not matter which angle you calculate.

Tip:
Do not approximate here but use the full display on the calculator in the next calculation.

Give answers to three significant figures unless requested otherwise.

1 Calculate *x*, correct to the nearest mm.

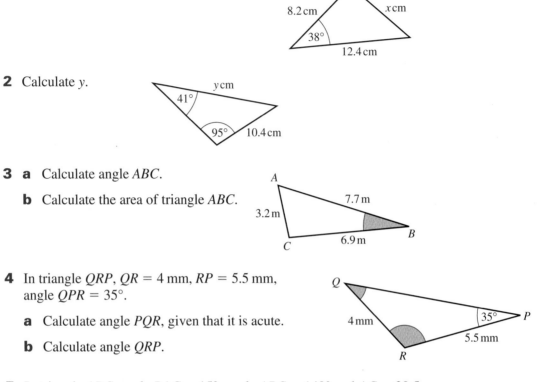

2 Calculate *y*.

3 a Calculate angle *ABC*.

  b Calculate the area of triangle *ABC*.

4 In triangle *QRP*, *QR* = 4 mm, *RP* = 5.5 mm, angle *QPR* = 35°.

  a Calculate angle *PQR*, given that it is acute.

  b Calculate angle *QRP*.

5 In triangle *ABC*, angle *BAC* = 15°, angle *ABC* = 140° and *AC* = 20.5 cm.

  a Calculate length *BC*.   **b** Calculate the area of triangle *ABC*.

6 A triangle has two equal sides of length 6 cm and one of the angles in the triangle is 40°.

  a Sketch two possible triangles.

  b For each triangle, calculate
    i the area of the triangle,
    ii the length of the third side.

7 Calculate angle *QRS*.

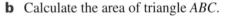
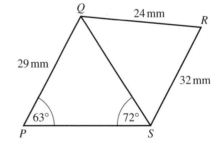

8 A circle, centre *O*, has radius 12 cm. Angle *OPQ* = 37°. Calculate

  a the length of the chord *PQ*,

  b the area of triangle *OPQ*.

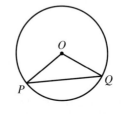

9 *ABCD* is a field in the shape of a quadrilateral.
AB = 25 m, *AD* = 15 m, *DC* = 17 m.
Angle *BAD* = 45°, angle *BCD* = 62°. Calculate

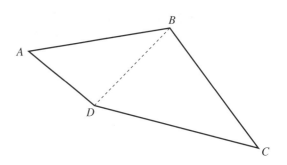

  a the length of the diagonal *BD*,

  b angle *DBC*,

  c angle *BDC*,

  d the area of the field.

10 In triangle *ABC*, *AB* = 8 cm and *AC* = 12 cm.

  a The area of the triangle is 24 cm². Given that angle *BAC* is acute, calculate angle *BAC* and length *BC*.

  b Show that when angle *BAC* is 150°, the area of triangle *ABC* is also 24 cm² and calculate the length *BC* in this case.

  c With the aid of diagrams, comment on your answers to parts **a** and **b**.

SKILLS CHECK **3A EXTRA** is on the CD

3.3 Degrees and radians

Degree and radian measure.

When the radius *OP* turns through an angle θ about *O*, the **sector** *POQ* is formed, with **arc length** *PQ*.

When the length of the arc is the same as the radius, the angle is 1 **radian** (1ᶜ).

Converting between radians and degrees

Make sure that you learn the following:

  π radians = 180°

To convert radians to degrees, multiply by $\dfrac{180}{\pi}$.

To convert degrees to radians, multiply by $\dfrac{\pi}{180}$.

Example 3.7 Write 40° in radians as a multiple of π.

Step 1: Multiply by $\dfrac{\pi}{180}$ $40° = 40 \times \dfrac{\pi}{180}$ radians $= \dfrac{2}{9}\pi$ radians.

and simplify, leaving π in your answer.

You will find it helpful to learn these common conversions:

Radians	Degrees
$\frac{1}{6}\pi$	30°
$\frac{1}{4}\pi$	45°
$\frac{1}{3}\pi$	60°
$\frac{1}{2}\pi$	90°
π	180°
2π	360°
1ᶜ	57° (to nearest degree)
0.017ᶜ (to 3 d.p.)	1°

299069

510 HIR

Arc length, area of sector.

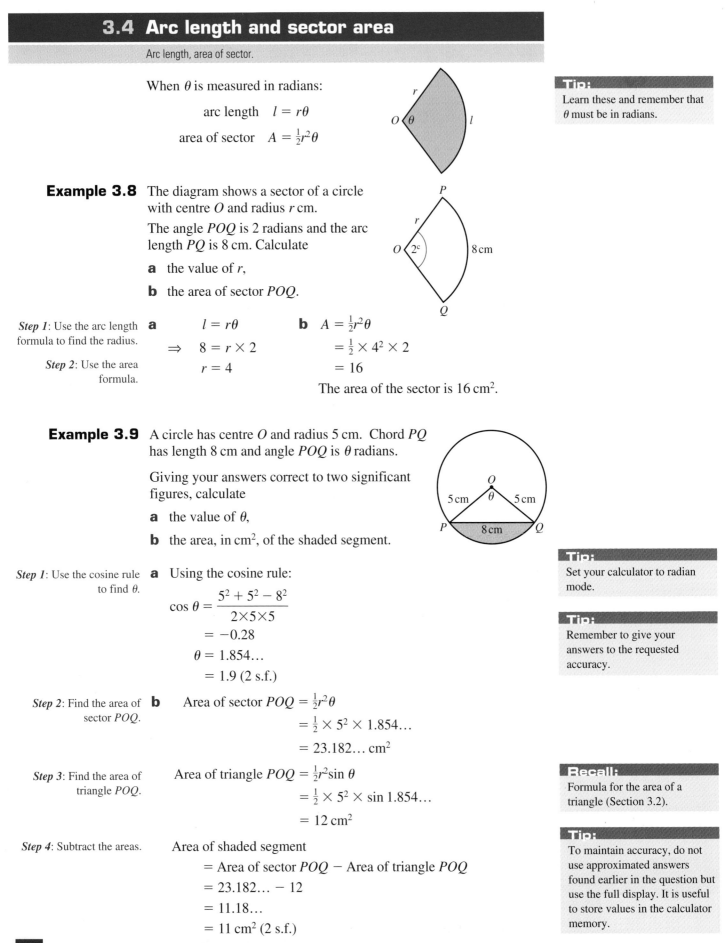

When θ is measured in radians:

arc length $l = r\theta$

area of sector $A = \frac{1}{2}r^2\theta$

Example 3.8   The diagram shows a sector of a circle with centre O and radius r cm.

The angle POQ is 2 radians and the arc length PQ is 8 cm. Calculate

a the value of r,

b the area of sector POQ.

Step 1: Use the arc length formula to find the radius.

Step 2: Use the area formula.

a $l = r\theta$ **b** $A = \frac{1}{2}r^2\theta$

$\Rightarrow$ $8 = r \times 2$ $= \frac{1}{2} \times 4^2 \times 2$

$r = 4$ $= 16$

The area of the sector is 16 cm².

Example 3.9   A circle has centre O and radius 5 cm. Chord PQ has length 8 cm and angle POQ is θ radians.

Giving your answers correct to two significant figures, calculate

a the value of θ,

b the area, in cm², of the shaded segment.

Step 1: Use the cosine rule to find θ.

a Using the cosine rule:

$$\cos\theta = \frac{5^2 + 5^2 - 8^2}{2 \times 5 \times 5}$$

$$= -0.28$$

$$\theta = 1.854\ldots$$

$$= 1.9 \ (2 \text{ s.f.})$$

Step 2: Find the area of sector POQ.

b   Area of sector $POQ = \frac{1}{2}r^2\theta$

$$= \frac{1}{2} \times 5^2 \times 1.854\ldots$$

$$= 23.182\ldots \text{ cm}^2$$

Step 3: Find the area of triangle POQ.

Area of triangle $POQ = \frac{1}{2}r^2\sin\theta$

$$= \frac{1}{2} \times 5^2 \times \sin 1.854\ldots$$

$$= 12 \text{ cm}^2$$

Step 4: Subtract the areas.

Area of shaded segment

$$= \text{Area of sector } POQ - \text{Area of triangle } POQ$$

$$= 23.182\ldots - 12$$

$$= 11.18\ldots$$

$$= 11 \text{ cm}^2 \ (2 \text{ s.f.})$$

Give answers to three significant figures unless requested otherwise.

1 Convert   **a** 280° to radians   **b** 1.5 radians to degrees.

2 Convert the following angles in radians to degrees.

   a $\frac{2}{3}\pi$ **b** $\frac{3}{4}\pi$ **c** $\frac{3}{2}\pi$ **d** $\frac{7}{12}\pi$

3 Convert these angles to radians, giving each angle in terms of π.

   a 45°   **b** 150°   **c** 330°   **d** 240°

4 *POQ* is a sector of a circle, centre *O*, radius 5 cm.
Angle *POQ* is 0.6 radians. Calculate

   a the length of the arc *PQ*,   **b** the perimeter of the sector,

   c the area of triangle *POQ*,   **d** the area of sector *POQ*.

5 *AOB* is a sector of a circle, centre *O*, radius 10.4 cm. The arc length *AB* is 12.48 cm.

   a Find angle *AOB*.   **b** Find the area of sector *AOB*.

6 A sector of a circle has area 27 cm² and radius 6 cm.

   a Calculate the angle of the sector, in radians.

   b Calculate the perimeter of the sector.

7 A circle, with centre *O*, has radius 8 cm.
A chord intersects the circle at *P* and *Q* and
angle *POQ* is θ radians, where θ is acute.

The area of triangle *POQ* is 24 cm². Find

   a the value of θ,

   b the area of sector *POQ*,

   c the area of the shaded segment.

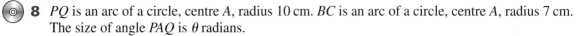

8 *PQ* is an arc of a circle, centre *A*, radius 10 cm. *BC* is an arc of a circle, centre *A*, radius 7 cm.
The size of angle *PAQ* is θ radians.

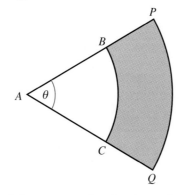

   a Find, in terms of θ, an expression for the perimeter of *BPQC*.

   b Given that the perimeter of *BPQC* is 14.5 cm, show that θ is 0.5.

   c Find, in cm², the area of *BPQC*.

9 In triangle ABC, $AB = 9$ cm, $AC = 6$ cm and angle $BAC = \frac{1}{6}\pi$ radians.
A circle, centre A, radius 2 cm, intersects the triangle at P and Q.
Calculate

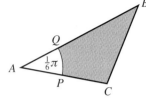

a length BC,　　　　　　　　　　　**b** arc length QP,

c the perimeter of the shaded region $QBCP$,　**d** the area of triangle ABC,

e the area of sector AQP,　　　　　　**f** the area of the shaded region $QBCP$.

10 A jigsaw piece is made from an equilateral triangle
ABC with sides of length 2 cm.
A sector of a circle, radius 0.5 cm, is cut away from each vertex.

a Find the perimeter of the jigsaw piece.

b Find the area of the jigsaw piece.

SKILLS CHECK **3B EXTRA** is on the CD

3.5 Trigonometric functions

Sine, cosine and tangent functions. Their graphs, symmetries and periodicity.

You will need to be able to recall the main features of the graphs of
$y = \sin x$, $y = \cos x$ and $y = \tan x$. Make sure that you can sketch
them for x in degrees or radians.

Remember that $180° = \pi$ radians.

Note:
sin is shorthand for sine, cos for cosine and tan for tangent.

Tip:
In module C2 you may use a graphical calculator to check.

$y = \sin x$

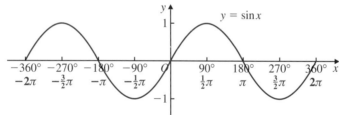

Note:
In degrees:
$\sin(x + 360°) = \sin x$.
In radians:
$\sin(x + 2\pi) = \sin x$.

The minimum value of $\sin x$ is -1 and the maximum value is 1.
The graph is periodic, repeating every $360°$ (2π radians).
The vertical line through every vertex (turning point) is an axis of
symmetry.

$y = \cos x$

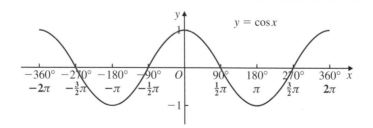

Recall:
$y = \cos x$ is a translation of
$y = \sin x$ by $90°$ ($\frac{1}{2}\pi$) to the left,
i.e. $\cos x = \sin(x + 90°)$,
or $\sin(x + \frac{1}{2}\pi)$ in radians.
(Section 1.2).

Note:
In degrees:
$\cos(x + 360°) = \cos x$.
In radians:
$\cos(x + 2\pi) = \cos x$.

The minimum value of $\cos x$ is -1 and the maximum value is 1.
The graph is periodic, repeating every $360°$ (2π radians).
The vertical line through every vertex (turning point) is an axis of
symmetry, in particular the y-axis.

$y = \tan x$

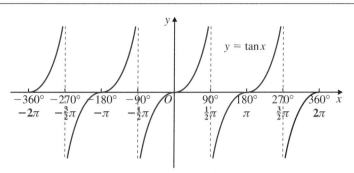

Note:
tan x takes every possible value in each 180° interval (π radians).

Note:
In degrees:
$\tan(x + 180°) = \tan x$.
In radians:
$\tan(x + \pi) = \tan x$.

Notice that $\tan x$ can take any value.
The graph is periodic, repeating every 180° (π radians).
There are no lines of symmetry.

The graph has **asymptotes** at $x = \pm 90°$ ($\pm\frac{1}{2}\pi$), $x = \pm 270°$ ($\pm\frac{3}{2}\pi$), and so on.

Note:
tan x is undefined at $\pm 90°$, $\pm 270°$, $\pm 450°$, …

Transformations of trigonometric graphs

You may be asked to sketch transformations of $y = \sin x$, $y = \cos x$ and $y = \tan x$, for example $y = \sin(x - 90°)$, $y = 2\cos x$, $y = \tan 2x$, $y = \cos x + 1$. For more on this, see Section 1.2.

Tip:
Make sure that you know how to do these.

Special angles

It is useful to recognise the sin, cos and tan of these special angles.

x	$\sin x$	$\cos x$	$\tan x$
30° ($\frac{1}{6}\pi$)	$\dfrac{1}{2}$	$\dfrac{\sqrt{3}}{2}$	$\dfrac{1}{\sqrt{3}}$
45° ($\frac{1}{4}\pi$)	$\dfrac{1}{\sqrt{2}}$	$\dfrac{1}{\sqrt{2}}$	1
60° ($\frac{1}{3}\pi$)	$\dfrac{\sqrt{3}}{2}$	$\dfrac{1}{2}$	$\sqrt{3}$

Note:
You will not be required to recognise these special angles in C2.

3.6 Trigonometric identities

Knowledge and use of $\tan\theta = \dfrac{\sin\theta}{\cos\theta}$ and $\sin^2\theta + \cos^2\theta = 1$.

Note:
In identities and equations, θ is often used for the angle.

The following **identities** are true for all values of θ.

$$\tan\theta = \frac{\sin\theta}{\cos\theta}$$

$$\sin^2\theta + \cos^2\theta = 1$$

Tip:
Learn these identities carefully; you will often need to apply them when solving trigonometric equations.

Example 3.10 Find the exact value of $\tan\theta$, given that $\sin\theta = -\frac{5}{13}$ and $\cos\theta = \frac{12}{13}$.

Step 1: Use the appropriate trig identity and simplify if necessary.

$$\tan\theta = \frac{\sin\theta}{\cos\theta} = \frac{-\frac{5}{13}}{\frac{12}{13}} = -\frac{5}{12}$$

Tip:
Notice the instruction to give the *exact* value. You need to work in fractions here.

Example 3.11 Express $\dfrac{5 + 4\sin^2\theta}{3 - 2\cos\theta}$ in the form $a + b\cos\theta$, where a and b are integers to be found.

Step 1: Write each expression in terms of $\cos\theta$.

$$5 + 4\sin^2\theta = 5 + 4(1 - \cos^2\theta)$$
$$= 5 + 4 - 4\cos^2\theta$$
$$= 9 - 4\cos^2\theta$$
$$= (3 - 2\cos\theta)(3 + 2\cos\theta)$$

Step 2: Simplify the given expression.

$$\frac{5 + 4\sin^2\theta}{3 - 2\cos\theta} = \frac{(3 - 2\cos\theta)(3 + 2\cos\theta)}{3 - 2\cos\theta}$$
$$= 3 + 2\cos\theta$$

Step 3: Compare coefficients.

Comparing with $a + b\cos\theta$ gives $a = 3$ and $b = 2$.

Tip:
Rearrange $\sin^2\theta + \cos^2\theta = 1$ to get $\sin^2\theta = 1 - \cos^2\theta$.

Recall:
The difference between two squares:
$p^2 - q^2 = (p - q)(p + q)$

3.7 Trigonometric equations

Solution of simple trigonometric equations in a given interval of degrees or radians.

The simplest trigonometric equations are of the form $\sin x = c$, $\cos x = c$ and $\tan x = c$, where c is a number.

Your calculator will give you *one* solution to an equation of this type, the **principal value** (PV). This lies in a particular range, depending on the function.

Tip:
To get the PV, key in $\sin^{-1}c$, $\cos^{-1}c$ or $\tan^{-1}c$ where c is a particular number. Make sure your calculator is in the correct mode: degrees or radians.

	In degrees	**In radians**
For sine function	$-90° \leqslant \text{PV} \leqslant 90°$	$-\frac{1}{2}\pi \leqslant \text{PV} \leqslant \frac{1}{2}\pi$
For cosine function	$0 \leqslant \text{PV} \leqslant 180°$	$0 \leqslant \text{PV} \leqslant \pi$
For tangent function	$-90° \leqslant \text{PV} \leqslant 90°$	$-\frac{1}{2}\pi \leqslant \text{PV} \leqslant \frac{1}{2}\pi$

You may be asked to find all the solutions in a given interval. To do this, find the principal value first. Then use the symmetries and periodicity of the graph to find further solutions in the the range.

You may be asked to give solutions in degrees or in radians. The following three examples are worked in degrees, with the answers in radians noted at the end of each part.

$\sin x = c$

Example 3.12 Find the values of x in the interval $0° \leqslant x \leqslant 360°$ for which

a $\sin x = 0.5$ **b** $\sin x = -0.9$

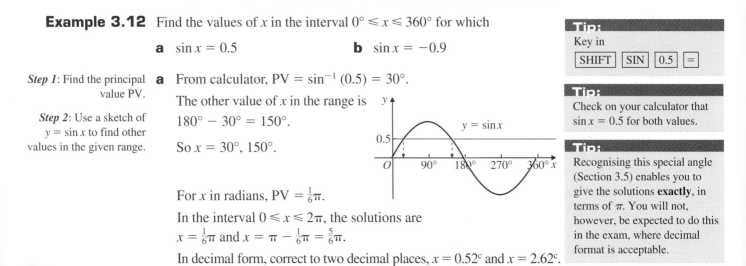

Step 1: Find the principal value PV.

Step 2: Use a sketch of $y = \sin x$ to find other values in the given range.

a From calculator, PV $= \sin^{-1}(0.5) = 30°$.

The other value of x in the range is $180° - 30° = 150°$.

So $x = 30°, 150°$.

For x in radians, PV $= \frac{1}{6}\pi$.

In the interval $0 \leqslant x \leqslant 2\pi$, the solutions are $x = \frac{1}{6}\pi$ and $x = \pi - \frac{1}{6}\pi = \frac{5}{6}\pi$.

In decimal form, correct to two decimal places, $x = 0.52^c$ and $x = 2.62^c$.

Tip:
Key in
| SHIFT | | SIN | | 0.5 | | = |

Tip:
Check on your calculator that $\sin x = 0.5$ for both values.

Tip:
Recognising this special angle (Section 3.5) enables you to give the solutions **exactly**, in terms of π. You will not, however, be expected to do this in the exam, where decimal format is acceptable.

b From calculator, PV = $\sin^{-1}(-0.9) = -64.15...°$

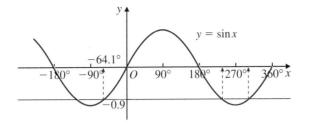
Recall:
For the sin function, the PV lies between $-90°$ and $90°$.

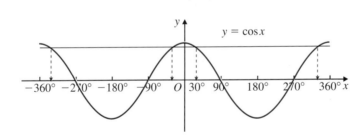

Other values of x *in range* are

$180° + 64.15...° = 244.15...°$

$360° - 64.15...° = 295.84...°$

So $x = 244°, 296°$ (nearest degree).

Tip:
Check on your calculator, but remember that you rounded, so you will not get $\sin x = -0.9$ exactly.

For x in radians, PV = -1.12^c. In the interval $0 \leqslant x \leqslant 2\pi$, the solutions are $x = \pi + 1.12^c$ and $x = 2\pi - 1.12^c$.

So $x = 4.26^c, 5.16^c$ (2 d.p.).

cos x = c

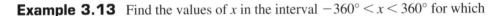

Example 3.13 Find the values of x in the interval $-360° < x < 360°$ for which

a $\cos x = \dfrac{\sqrt{3}}{2}$

b $\cos x = -0.8$

Step 1: Find the principal value PV.

a From calculator, PV = $\cos^{-1}\left(\dfrac{\sqrt{3}}{2}\right) = 30°$.

Step 2: Use a sketch of $y = \cos x$ to find other values in the given range.

From the graph, other values of x are

$360° - 30° = 330°, -30°$ and

$-360° + 30° = -330°$.

So $x = -330°, -30°, 30°, 330°$.

The solutions can be written $x = \pm 30°, \pm 330°$.

In radians: PV = $\frac{1}{6}\pi$.

Other solutions in the interval $-2\pi < x < 2\pi$ are $2\pi - \frac{1}{6}\pi$, $-\frac{1}{6}\pi$

and $-2\pi + \frac{1}{6}\pi$. So $x = \pm\frac{1}{6}\pi, \pm\frac{11}{6}\pi$.

In decimal form, correct to two decimal places, $x = \pm 0.52^c$ and $x = \pm 5.76^c$.

Tip:
Check these on your calculator.

Tip:
Since the y-axis is a line of symmetry, solutions will always be of the form $\pm$.

b From calculator, $PV = \cos^{-1}(-0.8) = 143.13...°$

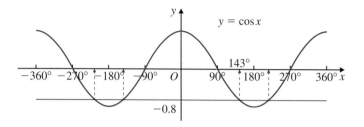

From the graph, other values of x are

$360° - 143.1...° = 216.8...°$,

$-143.13...°$ and $-216.8...°$.

So $x = \pm143°, \pm217°$ (nearest degree).

In radians: $PV = 2.5^c$ (1 d.p.).

Other solutions in the interval $-2\pi < x < 2\pi$ are $2\pi - 2.5^c, -2.5^c$ and $-2\pi + 2.5^c$. So $x = \pm2.5^c, \pm3.8^c$.

Tip:
Use the fact that the curve is symmetrical about the y-axis.

$\tan x = c$

Example 3.14 Find the values of x in the interval $-180° \le x \le 180°$ for which $\tan^2 x = 3$.

Step 1: Form equations in the form $\tan x = c$.

$\tan^2 x = 3 \Rightarrow \tan x = \sqrt{3}$ or $\tan x = -\sqrt{3}$

Consider $\tan x = \sqrt{3}$.

Step 2: For each equation, find the PV.

From calculator,
$PV = \tan^{-1}(\sqrt{3}) = 60°$.

Step 3: Use a sketch of $y = \tan x$ to find other values in the given range.

The other value of x in range is $-180° + 60° = -120°$.

So $x = -120°, 60°$.

Now consider $\tan x = -\sqrt{3}$.

From calculator,
$PV = \tan^{-1}(-\sqrt{3}) = -60°$.

The other value of x in range is $180° - 60° = 120°$.

So $x = -60°, 120°$.

The complete solution is $x = \pm60°, \pm120°$.

Tip:
There are two equations hidden in this question.

Note:
For the tan function, when you have found a solution in an interval of 180°, just add or subtract multiples of 180° to get further solutions.

In radians, where $-\pi < x < \pi$:

$\tan x = \sqrt{3}$ has $PV = \frac{1}{3}\pi$ and the other solution is $-\pi + \frac{1}{3}\pi = -\frac{2}{3}\pi$.

$\tan x = -\sqrt{3}$ has $PV = -\frac{1}{3}\pi$ and the other solution is $\pi - \frac{1}{3}\pi = \frac{2}{3}\pi$.

So the complete solution is $x = \pm\frac{1}{3}\pi, \pm\frac{2}{3}\pi$.

In decimal form, correct to two decimal places, $x = \pm1.05^c$ and $x = \pm2.09^c$.

Multiple angles

Take care if you are asked to solve equations involving multiples of x. This is illustrated in the following example.

Example 3.15 Find the values of x, in the interval $0° < x < 360°$, for which $\cos 2x = 0.5$.

Step 1: Make a substitution to get a simple equation.

Let $2x = \theta$, so the equation is $\cos \theta = 0.5$.

Step 2: Find the interval in which the new variable lies.

Interval required: $0° < x < 360°$

so $\qquad 0° < 2x < 720° \Rightarrow 0° < \theta < 720°$

Step 3: Solve the equation in the new variable.

For θ, PV $= \cos^{-1}(0.5) = 60°$.

Other solutions in interval $0° < \theta < 720°$ are

$360° - 60°, 360° + 60°, 720° - 60°$.

So $\theta = 60°, 300°, 420°, 660°$.

Step 4: Substitute back for x.

$2x = 60°, 300°, 420°, 660°$

$\Rightarrow x = 30°, 150°, 210°, 330°$

Using identities to solve trigonometric equations

Example 3.16 Find the values of θ, in the interval $-\pi < \theta < 2\pi$, for which $\sqrt{3} \sin \theta - \cos \theta = 0$.

Step 1: Rearrange to form an equation in $\tan \theta$, using an appropriate identity.

$\sqrt{3} \sin \theta - \cos \theta = 0$

$\sqrt{3} \sin \theta = \cos \theta$

($\div$ by $\cos \theta$) $\qquad \sqrt{3} \dfrac{\sin \theta}{\cos \theta} = 1$

$\sqrt{3} \tan \theta = 1$

($\div$ by $\sqrt{3}$) $\qquad \tan \theta = \dfrac{1}{\sqrt{3}}$

Step 2: Find the PV.

PV $= \tan^{-1}\left(\dfrac{1}{\sqrt{3}}\right) = \frac{1}{6}\pi$

Step 3: Use the periodicity of the tan curve to find other solutions in the given interval.

Other solutions in range are

PV $+ \pi = \frac{1}{6}\pi + \pi = \frac{7}{6}\pi$

PV $- \pi = \frac{1}{6}\pi - \pi = -\frac{5}{6}\pi$

In the interval $-\pi < \theta < 2\pi$, $\quad \theta = -\frac{5}{6}\pi, \frac{1}{6}\pi, \frac{7}{6}\pi$.

In decimal form, correct to two decimal places, $\theta = -2.62^c, 0.52^c, 3.67^c$.

Example 3.17 Find the values of θ in the interval $0 \leqslant \theta \leqslant 2\pi$ for which $2\sin^2 \theta = \sin \theta$, leaving your answers in terms of π.

Step 1: Form an equation $f(\theta) = 0$ and factorise if possible.

$2 \sin^2 \theta - \sin \theta = 0$

$\sin \theta (2 \sin \theta - 1) = 0$

Step 2: Solve the equations formed.

$\sin \theta = 0 \Rightarrow \theta = 0, \pi, 2\pi$

or $\qquad 2 \sin \theta - 1 = 0 \Rightarrow \sin \theta = \frac{1}{2}$

$\theta = \frac{1}{6}\pi, \frac{5}{6}\pi$

So $\theta = 0, \frac{1}{6}\pi, \frac{5}{6}\pi, \pi, 2\pi$.

In decimal form, correct to two decimal places where appropriate, $\theta = 0^c, 0.52^c, 2.62^c, 3.14^c, 6.28^c$.

Example 3.18 **a** Write the expression $3 - 2\sin^2\theta - 3\cos\theta$ in terms of $\cos\theta$.

b Hence solve $3 - 2\sin^2\theta - 3\cos\theta = 0$ for values of θ in the interval $0 \leqslant \theta \leqslant 2\pi$.

Step 1: Using an appropriate identity, form an equation in $\cos\theta$ only.

a $3 - 2\sin^2\theta - 3\cos\theta = 3 - 2(1 - \cos^2\theta) - 3\cos\theta$
$= 3 - 2 + 2\cos^2\theta - 3\cos\theta$
$= 2\cos^2\theta - 3\cos\theta + 1$

Tip:
Use $\sin^2\theta = 1 - \cos^2\theta$.

Step 2: Solve the equation in $\cos\theta$.

b $2\cos^2\theta - 3\cos\theta + 1 = 0$
$(\cos\theta - 1)(2\cos\theta - 1) = 0$
$\Rightarrow \qquad \cos\theta - 1 = 0$
$\cos\theta = 1$
$\theta = 0, 2\pi$

or $\qquad 2\cos\theta - 1 = 0$
$\cos\theta = \tfrac{1}{2}$
$\theta = \tfrac{1}{3}\pi, 2\pi - \tfrac{1}{3}\pi$
$= \tfrac{1}{3}\pi, \tfrac{5}{3}\pi$

Tip:
If the expression does not factorise, use the quadratic formula.

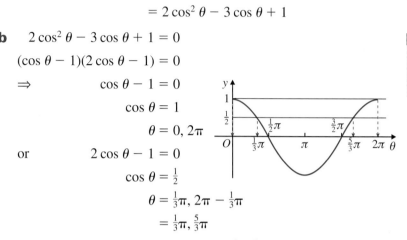

So, in the interval $0 \leqslant \theta \leqslant 2\pi$, $\theta = 0, \tfrac{1}{3}\pi, \tfrac{5}{3}\pi, 2\pi$.

In decimal form, correct to two decimal places where appropriate, $\theta = 0^c, 1.05^c, 5.24^c, 6.28^c$.

Example 3.19 Find the values of x in the interval $0° < x < 540°$ for which $\sin(x - 20°) = 0.2$. Give your answers correct to the nearest degree.

Step 1: Make a substitution to get a simple equation.

Let $x - 20° = \theta$.
The equation becomes $\sin\theta = 0.2$.

Tip:
Work out the interval for $x - 20°$.

Step 2: Find the interval in which the new variable lies.

Required interval: $\qquad 0° < x < 540°$
$(-20°)\qquad\qquad -20° < x - 20° < 520°$
Substitute θ $\qquad -20° < \theta < 520°$

Step 3: Work out the solutions for the new variable.

To solve $\sin\theta = 0.2$ for $-20° < \theta < 520°$, first find the PV.
$\mathrm{PV} = \sin^{-1}(0.2) = 11.5...°$

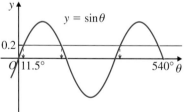

Other values of θ are

$180° - 11.5...° = 168.4...°$
$360° + 11.5...° = 371.5...°$
$540° - 11.5...° = 528.4...°$ (just out of range)

So $\theta = 11.5...°, 168.4...°, 371.5...°$.

Tip:
To maintain accuracy, work with uncorrected values if possible.

Tip:
Add 20° to each θ value.

Step 4: Write the solutions in terms of x.

$x - 20° = 11.5...°, 168.4...°, 371.5...°$
$x = 31.5...°, 188.4...°, 391.5...°$

Tip:
Remember to give your answers to the required accuracy and in the given range.

The values of x are $32°, 188°, 392°$ (correct to the nearest degree).

1 Solve the following equations for $0° \leqslant x < 360°$. If your answer is not exact, give it correct to the nearest degree.

    a $\sin x = 0.3$ **b** $\cos x = 0.5$ **c** $\tan x = -1.5$

    d $\cos 3x = \dfrac{\sqrt{3}}{2}$ **e** $\sin 2x = -0.5$ **f** $\tan (0.5x) = 1$

2 Solve the following equations for $0 \leqslant x \leqslant \pi$. You may give your answers exactly in terms of π, or in decimal form, correct to two decimal places.

> **Hint:**
> You may wish to use the table of special values on page 37.

    a $\sin x = \dfrac{\sqrt{3}}{2}$ **b** $\cos x = -\dfrac{1}{\sqrt{2}}$ **c** $\tan x = -\sqrt{3}$

    d $\sin 2x = -0.5$ **e** $\cos 3x = 0.5$ **f** $\tan 4x = \dfrac{1}{\sqrt{3}}$

3 Find all the values of x in the interval $-2\pi < x < 2\pi$ for which $\cos x = 0.75$, giving your answers in radians, correct to two decimal places.

4 Find all the values of x in the interval $-180° \leqslant x \leqslant 180°$ for which $2 \cos^2 x = \sqrt{3} \cos x$.

5 Find all the values of x in the interval $-360° \leqslant x \leqslant 360°$ for which $\sqrt{2} \sin^2 x - \sin x = 0$.

6 Show that $\dfrac{4 + \cos^2 \theta}{5 - \sin^2 \theta} = 1$, for all values of θ.

7 Find the exact value of $\tan \theta$, given that $\sin \theta = -\frac{4}{5}$ and $\cos \theta = -\frac{3}{5}$.

8 **a** Given that $\sin 3x = \cos 3x$, write down the value of $\tan 3x$.

    b Hence find all the solutions of the equation $\sin 3x = \cos 3x$ in the interval $0 < x < \pi$.

9 Find all the values of x in the interval $0 \leqslant x \leqslant 2\pi$ for which $2 \sin\left(x + \dfrac{\pi}{3}\right) = 1$.

10 **a** Given that $2 \sin^2 x = 1 - \cos x$, show that $2 \cos^2 x - \cos x - 1 = 0$.

    b Hence find all the values of x in the interval $0° \leqslant x \leqslant 360°$ for which $2 \sin^2 x = 1 - \cos x$.

    c **Write down** all the values of x in the interval $0° \leqslant x \leqslant 180°$ for which $2 \sin^2 2x = 1 - \cos 2x$.

SKILLS CHECK **3C EXTRA** is on the CD

Examination practice   Trigonometry

1 In triangle ABC, $AC = 50$ m, angle $BCA = 118°$ and angle $ABC = 35°$.

    a Calculate the length of AB, giving your answer to the nearest metre.

    b Calculate the area of triangle ABC.

2 To calculate the area of a field, *ABCD*, a farmer measures the boundary lengths and the length of a diagonal.

The measurements are

AB = 350 m

BC = 412 m

CD = 729 m

DA = 295 m

DB = 590 m.

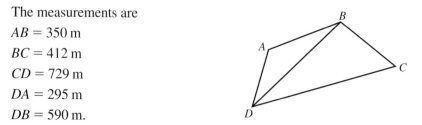

Calculate the area of the field, to the nearest hectare, where 1 hectare = 10 000 m².

3 In triangle *PQR*, angle *PQR* = 150° and *PQ* = 42 cm. The area of the triangle is 630 cm².

Calculate

a length *QR*,

b length *PR*, giving your answer to the nearest mm.

4 The diagrams show a square of side 6 cm and a sector of a circle of radius 6 cm and angle θ radians.

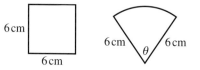

The area of the square is three times the area of the sector.

a Show that $\theta = \frac{2}{3}$.

b Show that the perimeter of the square is $1\frac{1}{2}$ times the perimeter of the sector.

[AQA (B) Nov 2002]

5 The diagram shows a circle with centre *O* and radius 3 cm. The points *A* and *B* on the circle are such that the angle *AOB* is 1.5 radians.

a Find the length of the minor arc *AB*.

b Find the area of the minor sector *OAB*.

c Show that the area of the shaded segment is approximately 2.3 cm².

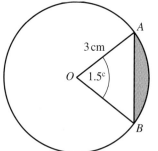

[AQA (A) Jan 2001]

6 The diagram shows a sector of a circle, with centre *O* and radius 6 cm. The midpoint of the chord *PQ* is *M*, and the angle $POM = \frac{\pi}{6}$ radians.

a Write down the exact values of:

   i the lengths of *PM* and *OM*;

   ii the length of the arc *PQ*;

   iii the area of the sector *POQ*.

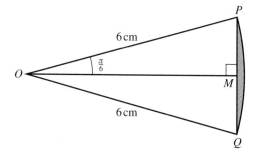

b Use the appropriate answers from part **a** to show that the area of the region shaded in the diagram is

$$m(2\pi - 3\sqrt{3}) \text{ cm}^2,$$

for some integer *m* whose value is to be determined.

[AQA (B) Jan 2001]

7 Solve the equation $\sin(x + 20°) = 0.5$ giving all solutions in the interval $0° < x < 360°$.

8 Solve the equation $\cos\left(x + \dfrac{\pi}{6}\right) = -0.5$, in the interval $0 < x < 2\pi$, giving your answers correct to two decimal places.

9 Solve the equation $\tan x = -\sqrt{3}$ in the interval $0 < x < 2\pi$, giving your answers in radians, correct to two decimal places.

10 a Given that $3 \cos 5x = 4 \sin 5x$, write down the value of $\tan 5x$.

b Hence, find all solutions of the equation

$$3 \cos 5x = 4 \sin 5x$$

in the interval $0° \leqslant x \leqslant 90°$, giving your answers correct to the nearest $0.1°$.

[AQA (B) Jan 2001]

11 It is given that x satisfies the equation

$$2 \cos^2 x = 2 + \sin x.$$

a Use an appropriate trigonometrical identity to show that

$$2 \sin^2 x + \sin x = 0.$$

b Solve this quadratic equation and hence find all the possible values of x in the interval $0 \leqslant x < 2\pi$.

[AQA (A) June 2003]

12 a i Express $\sin^2 x$ in terms of $\cos x$.

ii By writing $\cos x = y$, show that the equation $7 \cos x + 2 - 4 \sin^2 x = 0$ is equivalent to $4y^2 + 7y - 2 = 0$

b Solve the equation $4y^2 + 7y - 2 = 0$.

c Hence, solve the equation $7 \cos x + 2 - 4 \sin^2 x = 0$ giving all solutions to the nearest $0.1°$ in the interval $0° < x < 360°$.

[AQA (B) Jan 2003]

13 The angle θ radians, where $0 \leqslant \theta \leqslant 2\pi$, satisfies the equation $3 \tan \theta = 2 \cos \theta$.

a Show that $3 \sin \theta = 2 \cos^2 \theta$.

b Hence use an appropriate identity to show that $2 \sin^2 \theta + 3 \sin \theta - 2 = 0$.

c i Solve the quadratic equation in part **b**. Hence explain why the only possible value of $\sin \theta$ which will satisfy it is $\frac{1}{2}$.

ii Write down the values of θ for which $\sin \theta = \frac{1}{2}$ and $0 \leqslant \theta \leqslant 2\pi$

[AQA (A) Nov 2002]

14 Solve the equation $\cos(x - 20°) = -0.2$ giving all solutions to the nearest $0.1°$ in the interval $0° < x < 360°$.

15 The acute angle θ radians is such that

$$\sin \theta = \tfrac{5}{13}.$$

  a i Show that $\cos \theta = \tfrac{12}{13}$.

   ii Find the value of $\tan \theta$, giving your answer as a fraction.

  b Use your calculator to find the value of θ, giving your answer to three decimal places.

  c The diagram shows a sector of a circle of radius r cm and angle θ radians. The length of the arc which forms part of the boundary of the sector is 5 cm.

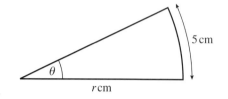

5 cm

r cm

   i Show that $r \approx 12.7$.

   ii Find the area of the sector, giving your answer to the nearest square centimetre.

[AQA (A) Jan 2003]

16 a The diagram shows an equilateral triangle ABC with sides of length 6 cm and an arc BC of a circle with centre A.

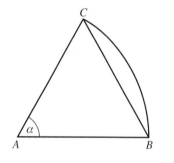

C

α

A B

   i Write down, in radians, the value of the angle α.

   ii Find the length of the arc BC.

   iii Show that the area of the triangle ABC is $9\sqrt{3}$ cm².

   iv Show that the area of the sector ABC is 6π cm².

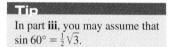

Tip

In part **iii**, you may assume that $\sin 60° = \tfrac{1}{2}\sqrt{3}$.

  b

C

A B

The diagram shows an ornament made from a flat sheet of metal. Its boundary consists of three arcs of circles. The straight lines AB, AC and BC are each of length 6 cm. The arcs BC, AC and AB have centres A, B and C respectively.

   i The boundary of the ornament is decorated with gilt edging. Find the total length of the boundary, giving your answer to the nearest centimetre.

   ii Find the area of one side of the ornament, giving your answer to the nearest square centimetre.

[AQA (A) June 2002]

4 Exponentials and logarithms

y = aˣ and its graph.

The graph of $y = a^x$ is called an **exponential** curve. When $x = 0$, $y = a^0 = 1$, so the curve goes through $(0, 1)$.

The shape of the curve depends on the value of a.

When $a > 1$:

As $x \to \infty$, $y \to \infty$.

As $x \to -\infty$, $y \to 0$.

Recall:
An exponent is another name for an index or a power (Section 1.1).

Note:
The y-value is always positive.

When $0 < a < 1$:

To get an idea of the general shape, let $a = \frac{1}{2}$ and consider $y = \left(\frac{1}{2}\right)^x$.

As $x \to \infty$, $y \to 0$.

As $x \to -\infty$, $y \to \infty$.

$\left(\frac{1}{2}\right)^x = 2^{-x}$, so when $a = \frac{1}{2}$, the curve is the graph of $y = 2^{-x}$. This is a reflection in the y-axis of $y = 2^x$.

Note:
Try substituting very large and very small numbers for x.

Recall:
Example 1.7.

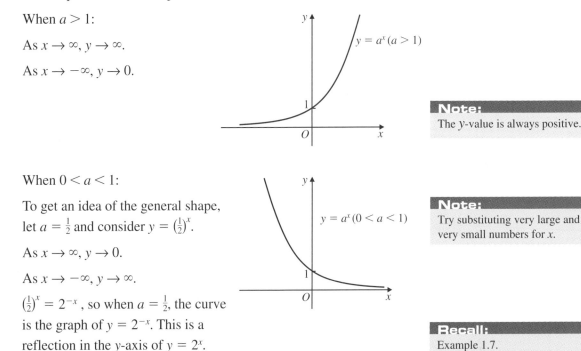

Transformations of exponential curves

You may be asked to transform exponential curves. See Section 1.2 for more on this.

Logarithms and the laws of logarithms.

The **logarithm** (log) of a positive number to a given **base** is the **power** to which the base must be raised to equal the number.

$$y = a^x \Leftrightarrow x = \log_a y$$

The base a is such that $a > 0$ and $a \neq 1$.
For example:

$$5^3 = 125 \Leftrightarrow \log_5 125 = 3 \quad \text{(5 is the base)}$$

$$10^2 = 100 \Leftrightarrow \log_{10} 100 = 2 \quad \text{(10 is the base)}$$

Note:
$\Leftrightarrow$ indicates a two-way implication: $y = a^x \Rightarrow x = \log_a y$ and $x = \log_a y \Rightarrow y = a^x$.

Example 4.1   Find x in each of the following.

    a $\log_3 81 = x$ **b** $\log_3 1 = x$

    c $\log_3 3 = x$ **d** $\log_3 \left(\frac{1}{9}\right) = x$

Step 1: Write in index form.   **a** $\log_3 81 = x \Rightarrow 3^x = 81$

Step 2: Solve the equation.   By inspection $x = 4$, since $3^4 = 81$.

    b $\log_3 1 = x \Rightarrow 3^x = 1$

        By inspection $x = 0$, since $3^0 = 1$.

> **Note:**
> For any base a, $\log_a 1 = 0$.

    c $\log_3 3 = x \Rightarrow 3^x = 3$

        By inspection $x = 1$, since $3^1 = 3$.

    d $\log_3 \left(\frac{1}{9}\right) = x \Rightarrow 3^x = \frac{1}{9}$

$$3^x = 3^{-2}$$
$$x = -2$$

> **Recall:**
> $\dfrac{1}{9} = \dfrac{1}{3^2} = 3^{-2}$
> (see Section 1.1).

Laws of logarithms

For any base a: Examples:

$$\log_a x + \log_a y = \log_a (xy) \qquad \log_a 3 + \log_a 4 = \log_a 12$$

$$\log_a x - \log_a y = \log_a \left(\frac{x}{y}\right) \qquad \log_a 20 - \log_a 5 = \log_a 4$$

$$k \log_a x = \log_a (x^k) \qquad 2\log_a 3 = \log_a 9$$

> **Tip:**
> Learn these laws and apply them accurately.

Special cases: Examples:

$$\log_a a = 1 \quad (\text{since } a^1 = a) \qquad \log_2 2 = 1$$

$$\log_a 1 = 0 \quad (\text{since } a^0 = 1) \qquad \log_3 1 = 0$$

$$\log_a (a^x) = x \log_a a = x \qquad \log_3 (3^4) = 4$$

$$\log_a \left(\frac{1}{x}\right) = \log_a (x^{-1}) = -\log_a x \qquad \log_a \left(\tfrac{1}{3}\right) = -\log_a 3$$

Example 4.2   Find the value of x.

    a $\log_a x = \log_a 20 - \log_a 15 + \log_a 3$

    b $\log_a x = 2\log_a 8 - 3\log_a 4$

Step 1: Simplify using the log laws.   **a** $\log_a x = \log_a \left(\dfrac{20 \times 3}{15}\right) = \log_a 4$

$$\Rightarrow x = 4$$

    b $\log_a x = 2\log_a 8 - 3\log_a 4$

$$= \log_a 8^2 - \log_a 4^3$$
$$= \log_a 64 - \log_a 64$$
$$= 0$$
$$\Rightarrow x = a^0 = 1$$

Example 4.3   Simplify $\log_a (a\sqrt{a})$.

$$\log_a(a\sqrt{a}) = \log_a a + \log_a (a^{\frac{1}{2}})$$
$$= 1 + \tfrac{1}{2} \log_a a$$
$$= 1 + \tfrac{1}{2}$$
$$= 1\tfrac{1}{2}$$

> **Tip:**
> You could write $a\sqrt{a} = a^{\frac{3}{2}}$; then $\log_a(a^{\frac{3}{2}}) = \tfrac{3}{2}\log_a a = \tfrac{3}{2}$.

Example 4.4 **a** Given $\log_2 16 = x$, find x.

b Write down the value of **i** $\log_2 16^3$ **ii** $\log_2 \dfrac{1}{16^2}$.

Step 1: Evaluate the log. **a** $\log_2 16 = x \Rightarrow 2^x = 16 \Rightarrow x = 4$

Step 2: Simplify, using the log laws. **b** **i** $\log_2 16^3 = 3(\log_2 16) = 3 \times 4 = 12$

ii $\log_2 \dfrac{1}{16^2} = \log_2 16^{-2} = -2(\log_2 16) = -2 \times 4 = -8$

Example 4.5 Given that $\log_3 x = \log_9 3$, find the value of x.

Step 1: Evaluate the right-hand side. Consider the right-hand side:

Let $\log_9 3 = y$

Then $9^y = 3$

$(3^2)^y = 3^1$

Equating indices:

$2y = 1$

$y = \frac{1}{2}$

Recall:
$(a^m)^n = a^{mn}$ (Section 1.1).

Step 2: Convert to exponential form. So $\log_3 x = \frac{1}{2}$

$\Rightarrow \qquad x = 3^{\frac{1}{2}} = \sqrt{3}$

The graph of $y = \log_a x$

When $x = 1$, $y = \log_a 1 = 0$, so the graph of $y = \log_a x$ goes through $(1, 0)$.

As $x \to 0$ (from the right), $y \to -\infty$.

As $x \to \infty$, $y \to \infty$.

It is interesting to note that the curves $y = \log_a x$ and $y = a^x$ are reflections of each other in the line $y = x$.

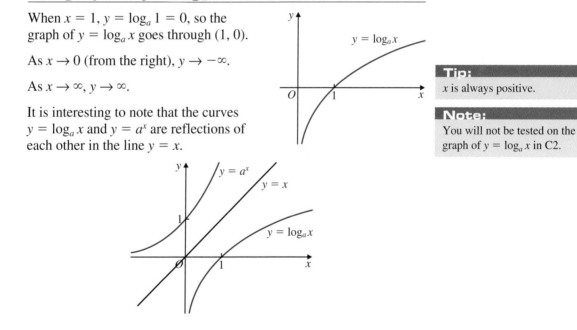

Tip:
x is always positive.

Note:
You will not be tested on the graph of $y = \log_a x$ in C2.

The solution of equations of the form $a^x = b$.

One way of solving equations of the form $a^x = b$ is to take logs to the base 10 of both sides.

Example 4.6 Solve **a** $5^x = 51$ **b** $2^{3x+1} = 12$.

Step 1: Take logs to base 10 of both sides.

Step 2: Simplify using the log laws.

Step 3. Solve the equation in x.

a $\log_{10}(5^x) = \log_{10} 51$

$x \log_{10} 5 = \log_{10} 51$

$$x = \frac{\log_{10} 51}{\log_{10} 5}$$

$$= 2.44 \ (3 \text{ s.f.})$$

b $\log_{10}(2^{3x+1}) = \log_{10} 12$

$(3x + 1)\log_{10} 2 = \log_{10} 12$

$$3x + 1 = \frac{\log_{10} 12}{\log_{10} 2}$$

$$= 3.584...$$

$$3x = 3.584... - 1 = 2.584...$$

$$x = 0.862 \ (3 \text{ s.f.})$$

Note:
Use $\boxed{\log}$ on your calculator. This is programmed to give logs to base 10. You could also use $\boxed{\ln}$ which gives logs to the base e. You will learn more about $\log_e x$ in C3.

Tip:
These logs are divided, not subtracted so do not try to cancel here.

SKILLS CHECK   **4A: Exponentials and logarithms**

1 Evaluate   **a** $\log_4 64$ **b** $\log_5 25$ **c** $\log_2 8$ **d** $\log_{16} 4$.

2 Evaluate   **a** $\log_2 8^3$ **b** $\log_3\left(\frac{1}{9}\right)$ **c** $\log_4 \sqrt{64}$ **d** $\dfrac{\log_3 27}{\log_3 9}$.

3 Find the value of x.

   a $\log_a x = \log_a 30 - \log_a 5 - \log_a 3$ **b** $\log_a x = 2\log_a 2 + 2\log_a 3$

4 Simplify   **a** $\log_a a^5$ **b** $\log_a\left(\dfrac{1}{\sqrt{a}}\right)$ **c** $4\log_a 1 + 3\log_a a$.

5 Express $\log_2 \sqrt{\dfrac{p^2 q}{2r^3}}$ in terms of $\log_2 p$, $\log_2 q$ and $\log_2 r$.

6 Given that $\log_a x = 2(\log_a 3 + \log_a 2)$, where a is a positive constant, find x.

7 **a** Write down the value of **i** $\log_3 3$ **ii** $\log_3 27$.

   b Find the value of $\log_3 2 - \log_3 54$.

8 Solve the following, giving your answers correct to three significant figures.

   a $2^x = 27$ **b** $3^{5x-2} = 20$ **c** $2\log_x 5 = 3$

9 **a** Show that $\log_a b + \log_a b^2 + \log_a b^3 + \cdots$ $(b \neq 1)$ is an arithmetic series and state the common difference of the series.

   b The sum of the first ten terms of the series is $k \log_a b$. Find k.

10 It is given that $3^{2x} = 10(3^x) - 9$.

   a Writing $y = 3^x$, show that $y^2 - 10y + 9 = 0$.

   b Solve $3^{2x} = 10(3^x) - 9$.

SKILLS CHECK   **4A EXTRA is on the CD**

1 a Write down the value of $\log_2 8$.

b Express $\log_2 9$ in the form $n\log_2 3$.

c Hence show that $\log_2 72 = m + n \log_2 3$, where m and n are integers.

[AQA (A) Nov 2002]

2 a Given that $\log_a x = \log_a 5 + 2 \log_a 3$, where a is a positive constant, show that $x = 45$.

b i Write down the value of $\log_2 2$.

ii Given that $\log_2 y = \log_4 2$, find the value of y.

[AQA (B) Jan 2002]

3 a Given that $\log_a x = 2(\log_a k - \log_a 2)$, where a is a positive constant, show that $k^2 = 4x$.

b Given that $\log_3 y = \log_9 27$, find the value of y.

[AQA (B) Jan 2003]

4 a Show that $\log_2 8 = 3$.

b Find the value of **i** $\log_2 (8^4)$ **ii** $\log_2 \dfrac{1}{\sqrt{8}}$.

[AQA (A) Jan 2002]

5 a Write down the value of **i** $\log_2 2$ **ii** $\log_2 8$.

b Find the value of $\log_2 3 - \log_2 24$.

[AQA (B) Nov 2003]

6 a Find the value of $\log_4 64$.

b Given that $\log_x 27 = \log_4 64$, find the value of x.

7 Find the value of x if $3^x = 60$, giving your answer correct to two decimal places.

8 Find the value of x if $3^{x-2} = 2^{x+1}$. Give your answer correct to one decimal place.

9 a Given that $2 + \log_3 x = \log_3 y$, show that $y = 9x$.

b Hence, or otherwise, solve $2 + \log_3 x = \log_3 (5x + 2)$.

10 a Express 4^{2x+1} as a power of 2.

b The variable x satisfies the equation $5^x \times 4^{2x+1} = 2^{x+3}$. Show that $x = \dfrac{\log_{10} 2}{\log_{10} 40}$.

5 Differentiation

5.1 Differentiation of rational powers of *x*

Differentiation of x^n, where *n* is a rational number, and related sums and differences.

In module C1 you differentiated ax^n, where *n* is a positive integer. In C2, the theory is extended to include the case when *n* is any rational number.

Note:
n can be positive or negative, an integer or a fraction.

For any rational number *n*,

$$y = ax^n \Rightarrow \frac{dy}{dx} = anx^{n-1}$$

Recall:
Multiply by the power of *x* and decrease the power by 1.

You also need to remember the following:

$$y = f(x) \pm g(x) \Rightarrow \frac{dy}{dx} = f'(x) \pm g'(x)$$

Recall:
This means that you can differentiate term by term.

Example 5.1 Differentiate with respect to *x*:

a $y = \dfrac{1}{x^2}$ **b** $y = 3\sqrt{x} + \dfrac{2}{x^4}$

Step 1: If necessary, write in index form first.

a $y = \dfrac{1}{x^2} = x^{-2}$

Step 2: Differentiate term by term using the rule.

$\dfrac{dy}{dx} = -2x^{-3}$

Recall:
$\sqrt[m]{x} = x^{\frac{1}{m}}$ and $x^{-n} = \dfrac{1}{x^n}$
(Section 1.1).

b $y = 3\sqrt{x} + \dfrac{2}{x^4} = 3x^{\frac{1}{2}} + 2x^{-4}$

$\dfrac{dy}{dx} = 3 \times \frac{1}{2}x^{-\frac{1}{2}} + 2 \times (-4)x^{-5}$

$= \frac{3}{2}x^{-\frac{1}{2}} - 8x^{-5}$

$= \dfrac{3}{2\sqrt{x}} - \dfrac{8}{x^5}$

Note:
You could leave the answer in index form.

Example 5.2 A curve has equation $y = 2\sqrt{x} - 2x$.

Differentiate *y* with respect to *x* and hence find the gradient at the point $(4, -4)$.

Step 1: Write in index form.

$y = 2\sqrt{x} - 2x = 2x^{\frac{1}{2}} - 2x$

Step 2: Differentiate term by term.

$\dfrac{dy}{dx} = 2 \times \frac{1}{2}x^{-\frac{1}{2}} - 2$

$= x^{-\frac{1}{2}} - 2$

$= \dfrac{1}{\sqrt{x}} - 2$

Step 3: Substitute the *x*-value into the derivative.

When $x = 4$, $\dfrac{dy}{dx} = \dfrac{1}{\sqrt{4}} - 2 = -\frac{3}{2}$.

The gradient at $(4, -4)$ is $-\frac{3}{2}$.

Recall:
To find the gradient at a point, substitute the *x*-value into
$\dfrac{dy}{dx}$ (C1 Section 3.1).

Example 5.3 It is given that $f(x) = \left(\dfrac{1}{x} - \dfrac{1}{x^2}\right)^2$.

    a Find $f'(x)$.

    b Find $f'(1)$ and hence write down the gradient of the curve $y = f(x)$ at the point $(1, 0)$.

Step 1: Write in index form.

a $f(x) = \left(\dfrac{1}{x} - \dfrac{1}{x^2}\right)^2$

$\qquad = (x^{-1} - x^{-2})^2$

$\qquad = x^{-2} - 2x^{-3} + x^{-4}$

Step 2: Differentiate term by term.

$f'(x) = -2x^{-3} - 2(-3)x^{-4} + (-4)x^{-5}$

$\qquad = -2x^{-3} + 6x^{-4} - 4x^{-5}$

Step 3: Substitute the x-value into the derivative.

b $f'(1) = -2(1)^{-3} + 6(1)^{-4} - 4(1)^{-5}$

$\qquad\quad = -2 + 6 - 4$

$\qquad\quad = 0$

The gradient at $(1, 0)$ is given by $f'(1)$.

Hence the gradient at $(1, 0)$ is zero.

> **Recall:**
> $(a - b)^2 = a^2 - 2ab + b^2$

> **Recall:**
> Index laws (Section 1.1).

> **Note:**
> You could write
> $f'(x) = -\dfrac{2}{x^3} + \dfrac{6}{x^4} - \dfrac{4}{x^5}$

Example 5.4 Differentiate $y = \dfrac{6x^2 + x^3 - 2x}{2x}$ with respect to x.

Step 1: Simplify to obtain terms in index form.

$y = \dfrac{6x^2 + x^3 - 2x}{2x}$

$\quad = \dfrac{6x^2}{2x} + \dfrac{x^3}{2x} - \dfrac{2x}{2x}$

$\quad = 3x + \tfrac{1}{2}x^2 - 1$

Step 2: Differentiate term by term.

$\dfrac{dy}{dx} = 3 + x$

> **Tip:**
> Divide each term in the numerator by $2x$, then simplify each term individually.

> **Recall:**
> $\dfrac{x^p}{x^q} = x^{p-q}$ (Section 1.1).

Example 5.5 Given that $y = \dfrac{x^2 + x}{\sqrt{x}}$, show that $\dfrac{dy}{dx} = 3x + \dfrac{1}{2\sqrt{x}}$.

Step 1: Simplify to obtain terms in index form.

$y = x^2 + \dfrac{x}{\sqrt{x}}$

$\quad = \dfrac{x^2}{\sqrt{x}} + \dfrac{x}{\sqrt{x}}$

$\quad = \dfrac{x^2}{x^{\frac{1}{2}}} + \dfrac{x}{x^{\frac{1}{2}}}$

$\quad = x^{\frac{3}{2}} + x^{\frac{1}{2}}$

Step 2: Differentiate term by term.

$\dfrac{dy}{dx} = \tfrac{3}{2}x^{\frac{1}{2}} + \tfrac{1}{2}x^{-\frac{1}{2}}$

Step 3: Simplify to the required form.

$\qquad = \dfrac{3\sqrt{x}}{2} + \dfrac{1}{2\sqrt{x}}$

$\qquad = \dfrac{3\sqrt{x}}{2} \times \dfrac{\sqrt{x}}{\sqrt{x}} + \dfrac{1}{2\sqrt{x}}$

$\qquad = \dfrac{3x}{2\sqrt{x}} + \dfrac{1}{2\sqrt{x}}$

$\qquad = 3x + \dfrac{1}{2\sqrt{x}}$

> **Tip:**
> Divide each term in the numerator by $\sqrt{x}$.

> **Tip:**
> It is often easier to simplify terms involving square roots and negative indices when they are not written in index form.

1 Differentiate the following with respect to x.

   a $y = x^{-3}$ **b** $y = \dfrac{2}{x^5}$

   c $y = 4x^{\frac{5}{2}}$ **d** $y = \dfrac{6}{\sqrt{x}}$

2 It is given that $f(x) = x^2\left(\dfrac{1}{x^4} + \dfrac{2}{x^2}\right)$. Find $f'(x)$.

3 It is given that $y = \dfrac{x^2 - x^{\frac{3}{2}}}{2x^2}$.

   a Find $\dfrac{dy}{dx}$.

   b Find the gradient of the curve at $x = 1$.

4 Evaluate $f'(4)$, where $f(x) = (2x - 3\sqrt{x})^2$.

5 The sketch shows the curve $y = \dfrac{3}{2x^2}$.

The points A, B and C on the curve have x-coordinates -2, -1 and 3 respectively.

Find the gradient of the curve at each of the points A, B and C.

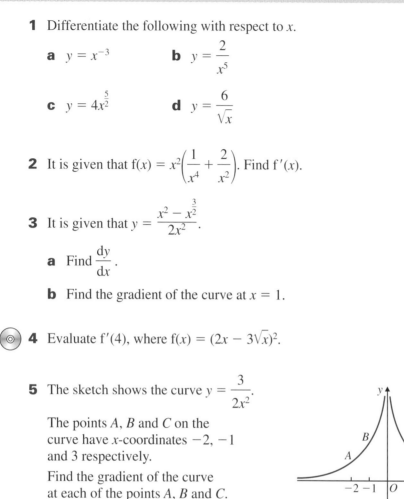

6 Find the values of x for which the gradient of the curve $y = x^2 + 16x^{-2}$ is zero.

7 Find the values of x for which the gradient of the curve $y = \dfrac{3}{x} + 9x$ is the same as the gradient of the line $y = 3x + 1$.

8 Find the gradient of the curve $y = 2\sqrt{x^3} - 4x$ at the origin.

9 Given that $y = \dfrac{x^3 - 5x}{\sqrt{x}}$, show that $\dfrac{dy}{dx} = \dfrac{5(x^2 - 1)}{2\sqrt{x}}$.

10 Given that $y = x^3 + \dfrac{3}{x^3}$, find the value of $\dfrac{d^2y}{dx^2}$ when $x = 1$.

Hint: differentiate $\dfrac{dy}{dx}$ to get $\dfrac{d^2y}{dx^2}$ (C1 Section 3.6).

5.2 Equations of tangents and normals

In C1 you found the **equation of the tangent** and the **equation of the normal** to a curve at a given point. You need to be able to apply all the same theory in C2.

Recall:
C1 Section 3.4.

Here are some reminders.

The **tangent** at a point $P(x_1, y_1)$ is the line that touches the curve at P.

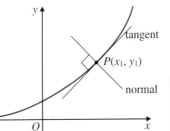

The **normal** at P is the line through P perpendicular to the tangent at P.

If m_1 is the gradient of the tangent at P and m_2 is the gradient of the normal at P, then $m_1 \times m_2 = -1$.

Recall:
The product of the gradients of perpendicular lines is -1 (C1 Section 2.2).

The equation of a straight line, with gradient m, through the point (x_1, y_1) can be written $y - y_1 = m(x - x_1)$.

Recall:
Equation of a line (C1 Section 2.1).

Example 5.6 A curve has equation $y = \dfrac{4}{x^3} - \dfrac{x^2}{4}$ and the point $P(2, -\tfrac{1}{2})$ lies on the curve.

a The equation of the tangent at P is $px + qy = r$, where p, q and r are integers. Find the values of p, q and r.

b Find an equation of the normal at P.

Step 1: Differentiate to find the gradient function.

a $y = \dfrac{4}{x^3} - \dfrac{x^2}{4} = 4x^{-3} - \tfrac{1}{4}x^2$

$\dfrac{\mathrm{d}y}{\mathrm{d}x} = -12x^{-4} - \tfrac{1}{2}x$

Tip:
Write all terms in index form before differentiating.

Step 2: Substitute the x-value to get the gradient of the tangent at P.

When $x = 2$, $\dfrac{\mathrm{d}y}{\mathrm{d}x} = -12 \times 2^{-4} - \tfrac{1}{2} \times 2 = -\tfrac{7}{4}$.

The gradient of the tangent at P is $-\tfrac{7}{4}$.

Tip:
Take care with negatives.

Step 3: Use $y - y_1 = m(x - x_1)$.

Equation of the tangent at P:

$y - (-\tfrac{1}{2}) = -\tfrac{7}{4}(x - 2)$

$y + \tfrac{1}{2} = -\tfrac{7}{4}(x - 2)$

Step 4: Rearrange to the required format and state the values of p, q and r.

$4(y + \tfrac{1}{2}) = -7(x - 2)$

$4y + 2 = -7x + 14$

$7x + 4y = 12$

So $p = 7$, $q = 4$, $r = 12$.

Tip:
Rearrange the equation to the given format.

Step 5: Find the gradient of the normal at P.

b At P, gradient of tangent $= -\tfrac{7}{4}$

$\Rightarrow$ gradient of normal $= \tfrac{4}{7}$

Tip:
$m_1 \times m_2 = -1$, so find the negative reciprocal.

Step 6: Use $y - y_1 = m(x - x_1)$.

Equation of normal at P:

$y - (-\tfrac{1}{2}) = \tfrac{4}{7}(x - 2)$

$y + \tfrac{1}{2} = \tfrac{4}{7}(x - 2)$

Tip:
The equation may be left in this format if one is not specified.

5.3 Stationary points

In C1 you found **stationary points** and investigated their **nature**. You need to be able to apply this theory in C2. Here are some reminders.

Recall:
At a stationary point, the gradient of the curve is zero (C1 Section 3.6).

At a stationary point, $\dfrac{dy}{dx} = 0$.

If $\dfrac{dy}{dx} = 0$ and $\dfrac{d^2y}{dx^2} < 0$, the stationary point is a maximum point.

If $\dfrac{dy}{dx} = 0$ and $\dfrac{d^2y}{dx^2} > 0$, the stationary point is a minimum point.

Recall:
Conditions for maximum and minimum (C1 Section 3.6).

Example 5.7 A curve has equation $y = \dfrac{1}{x} + 32x^2$. Find the coordinates of the stationary point on the curve and determine its nature.

Step 1: Write in index form.

$$y = \frac{1}{x} + 32x^2 = x^{-1} + 32x^2$$

Tip:
Write in index form before differentiating.

Step 2: Find $\dfrac{dy}{dx}$.

$$\frac{dy}{dx} = -x^{-2} + 64x = -\frac{1}{x^2} + 64x$$

Step 3: Set $\dfrac{dy}{dx} = 0$ and solve for x.

$$\frac{dy}{dx} = 0 \text{ when } -\frac{1}{x^2} + 64x = 0$$

$$64x = \frac{1}{x^2}$$

$$x^3 = \tfrac{1}{64}$$

$$x = \sqrt[3]{\tfrac{1}{64}} = \tfrac{1}{4}$$

Tip:
Multiply both sides by x^2 and divide by 64.

Step 4: Calculate the y-coordinate.

When $x = \tfrac{1}{4}$, $y = \dfrac{1}{x} + 32x^2 = \dfrac{1}{\frac{1}{4}} + 32 \times (\tfrac{1}{4})^2 = 6$.

Hence there is a stationary point at $(\tfrac{1}{4}, 6)$.

Step 5: Find $\dfrac{d^2y}{dx^2}$ and substitute the x-value.

$$\frac{dy}{dx} = -x^{-2} + 64x \Rightarrow \frac{d^2y}{dx^2} = 2x^{-3} + 64$$

When $x = \tfrac{1}{4}$, $\dfrac{d^2y}{dx^2} = 2 \times (\tfrac{1}{4})^{-3} + 64 = 192$.

Since $\dfrac{d^2y}{dx^2} > 0$, $(\tfrac{1}{4}, 6)$ is a minimum point.

Tip:
You must include sufficient working to show whether the second differential is positive or negative.

5.4 Application problems

In C1 you used differentiation to solve application problems. This is extended in C2 to include functions in rational powers of x. Problems may involve finding greatest or least values of a variable, given certain conditions.

Recall:
C1 Section 3.7.

Example 5.8 A cylindrical tin, closed at both ends, is made from thin sheet metal. The radius of the base of the cylinder is r cm and the volume of the tin is 1024π cm³.

a Show that the total surface area, S cm², of the cylinder is given by

$$S = \frac{2048\pi}{r} + 2\pi r^2.$$

b Find the value of r that gives a minimum total surface area and state the value of this surface area, in terms of π.

Step 1: Draw a diagram showing given information.

a Let the height be h cm.

Volume $= 1024\pi$

$\Rightarrow \pi r^2 h = 1024\pi$

Step 2: Write unknown measures in terms of the given variable.

$$h = \frac{1024}{r^2}$$

Tip:
Use the condition that the volume is 1024π to express h in terms of r.

Step 3: Form an expression in r for the surface area.

$S = 2\pi rh + 2\pi r^2$

$$= 2\pi r\left(\frac{1024}{r^2}\right) + 2\pi r^2$$

$$= \frac{2048\pi}{r} + 2\pi r^2$$

Tip:
Find the curved surface area and the area of the two circular ends.

Step 4: Find $\dfrac{dS}{dr}$.

b $S = 2048\pi r^{-1} + 2\pi r^2$

$$\frac{dS}{dr} = -2048\pi r^{-2} + 4\pi r$$

Tip:
Write all terms in index form before differentiating.

Step 5: Set $\dfrac{dS}{dr} = 0$ and solve for r.

$\dfrac{dS}{dr} = 0$ when $-2048\pi r^{-2} + 4\pi r = 0$

$$4\pi r = \frac{2048\pi}{r^2}$$

$(\times r^2)$ $\qquad\qquad\qquad 4\pi r^3 = 2048\pi$

$(\div 4\pi)$ $\qquad\qquad\qquad r^3 = \dfrac{2048\pi}{4\pi} = 512$

$$r = \sqrt[3]{512} = 8$$

There is a stationary value when $r = 8$.

$$\frac{dS}{dr} = -2048\pi r^{-2} + 4\pi r \Rightarrow \frac{d^2S}{dr^2} = 4096\pi r^{-3} + 4\pi$$

Step 6: Check the nature of the stationary value.

When $r = 8$, $\dfrac{d^2S}{dr^2} = 4096\pi \times 8^{-3} + 4\pi = 37.6... > 0$

So S has a minimum value when $r = 8$.

When $r = 8$, $S = \dfrac{2048\pi}{8} + 2\pi \times 8^2 = 384\pi$.

Tip:
Remember to give your answer in terms of π and include the units.

Step 7: Substitute for r into S.

The minimum surface area is 384π cm².

1 A curve has equation $y = 2x^{\frac{3}{2}} - 4x^{\frac{5}{2}} + 2x$.

   a Find $\dfrac{dy}{dx}$.

   b Show that the equation of the tangent to the curve at $(1, 0)$ is $y + 5x = 5$.

   c Find the equation of the normal at the point $(1, 0)$, writing your answer in the form $ax + by + c = 0$, where a, b and c are integers.

2 Find the equation of the normal to the curve $y = \dfrac{4}{x} + x^2$ at the point where $x = 1$, giving your answer in the form $ax + by + c = 0$, where a, b and c are integers.

3 Find the equation of the tangent to the curve $y = 40\sqrt{x}$ at the point where the gradient of the curve is 5, writing your answer in the form $y = mx + c$.

4 A curve has equation $y = \dfrac{x^2 - 2x^3}{x^5}$. The point $P(-1, -3)$ lies on the curve.

   a Find the equation of the tangent to the curve at P in the form $y = mx + c$.

   b Find the x-coordinate of the stationary point on the curve and determine the nature of the stationary point.

5 The diagram shows a sketch of the curve $y = 12x^{\frac{1}{2}} - 2x^{\frac{3}{2}}$.

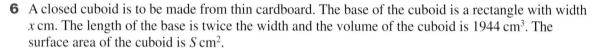

   a Find the x-coordinate of the maximum point B.

   b Hence state the maximum value of y, leaving your answer in the form $a\sqrt{2}$, where a is an integer to be found.

6 A closed cuboid is to be made from thin cardboard. The base of the cuboid is a rectangle with width x cm. The length of the base is twice the width and the volume of the cuboid is $1944\ \text{cm}^3$. The surface area of the cuboid is $S\ \text{cm}^2$.

   a Show that $S = 4x^2 + 5832x^{-1}$.

   b Given that x can vary, find the value of x that makes the surface area a minimum.

   c Find the minimum value of the surface area.

7 It is given that $f(x) = \dfrac{1}{\sqrt[3]{x}}$.

   a Show that $f'(8) = -\frac{1}{48}$.

   b Hence find the equation of the tangent to the curve $y = f(x)$ at the point where $x = 8$, giving your answer in the form $ax + by + c = 0$, where a, b and c are integers.

8 The diagram shows a prism where the cross-section is in the form of a sector OPQ of a circle, centre O and radius x cm. The length of the prism is $2x$ cm and the angle POQ is θ radians. The volume of the prism is $4608\ \text{cm}^3$.

   a Show that $\theta = \dfrac{4608}{x^3}$.

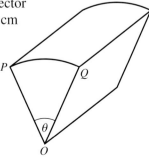

   b Show that the surface area, $S\ \text{cm}^2$, of the prism is given by $S = 4x^2 + \dfrac{13\,824}{x}$.

   c Given that x can vary, find the minimum value of the surface area of the prism, verifying that you have found the minimum value.

9 The function f is defined for $x > 0$ by $f(x) = -\dfrac{3}{2\sqrt{x}}$.

   a Find $f'(x)$.

   b Find the gradient of the normal when $x = 4$.

10 Find the equation of the tangent to the curve $y = x^2 - \dfrac{2}{x^2}$ at the point where $x = 2$.

   Give your answer in the form $ax + by + c = 0$, where a, b and c are integers and $a > 0$.

SKILLS CHECK **5B EXTRA** is on the CD

Examination practice   Differentiation

1 **a** Express $x^2\sqrt{x}$ in the form x^p.

   b Given that

$$y = x^2\sqrt{x},$$

   find the value of $\dfrac{dy}{dx}$ at the point where $x = 9$.   [AQA (A) June 2001]

2 The graph of

$$y = x + 4x^{-2}$$

has one stationary point.

   a Find $\dfrac{dy}{dx}$.

   b Find the coordinates of the stationary point.

   c Find the value of $\dfrac{d^2y}{dx^2}$ at the stationary point, and hence determine whether the stationary point is a maximum or a minimum.   [AQA (A) June 2003]

3 The curve with equation $y = 2x + \dfrac{27}{x^2} - 7$ is defined for $x > 0$, and is sketched below

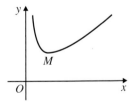

   i Find $\dfrac{dy}{dx}$.

   ii The curve has a minimum point M. Find the x-coordinate of M.   [AQA (B) Jan 2003]

4 The function f is defined for $x \geqslant 0$ by $f(x) = x^{\frac{1}{2}} + 2$.

   i Find $f'(x)$.

   ii Hence find the gradient of the curve $y = f(x)$ at the point for which $x = 4$.   [AQA (A) Jan 2003]

5 The curve with equation $y = 2x - x^{\frac{3}{2}}$ is defined for $x \geqslant 0$, and is sketched below.

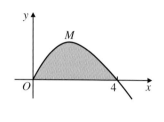

i Find $\dfrac{dy}{dx}$.

ii The curve has a maximum point M. Show that the x-coordinate of M is $\frac{16}{9}$.

[AQA (B) Nov 2003]

6 Given that $y = \dfrac{x - 2}{x^2}$,

i show that $\dfrac{dy}{dx} = \dfrac{4}{x^3} - \dfrac{1}{x^2}$;

ii show that y has a stationary value when $x = 4$;

iii find the value of $\dfrac{d^2y}{dx^2}$ when $x = 4$, and deduce the nature of this stationary value.

[AQA (B) June 2001]

7 A curve has equation $y = x^2 + \dfrac{81}{x^2}$. Its graph is sketched below.

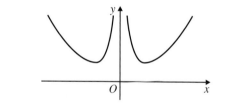

i Find $\dfrac{dy}{dx}$.

ii Show that the stationary points of the curve occur when $x^4 = 81$.

iii Hence find the x-coordinates of the stationary points.

iv Find the value of the y-coordinate at each stationary point.

[AQA (B) Jan 2004]

8 a It is given that $y = 2\pi x^2 + \dfrac{1000}{x}$.

i Find $\dfrac{dy}{dx}$.

ii Show that $\dfrac{dy}{dx} = 0$ when $x^3 = \dfrac{250}{\pi}$.

iii Find $\dfrac{d^2y}{dx^2}$.

iv Verify that $\dfrac{d^2y}{dx^2} = 12\pi$ when $x^3 = \dfrac{250}{\pi}$.

v Find, to one decimal place, the value of x for which y has a stationary value and state whether this stationary value is a maximum or a minimum.

b A closed cylindrical tin can contains 500 cm^3 of liquid when full. The can has base radius r cm and total external surface area A cm^2.

It is given that

$$A = 2\pi r^2 + \frac{1000}{r}.$$

Use your results from part **a** to find the smallest possible value for the total external surface area of the can. Give your answer to the nearest square centimetre. [AQA (A) Nov 2002]

9 A wire of length 10 cm is cut into two pieces. One of these pieces is bent to form an equilateral triangle of side x cm and the other piece is bent to form a sector of a circle of angle θ radians and radius x cm as shown below.

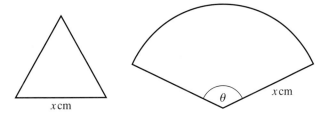

a Show that $5x + x\theta = 10$.

b The sum of the areas of the triangle and sector is denoted by A cm^2.

> **Tip:**
> $\sin \dfrac{\pi}{3} = \dfrac{\sqrt{3}}{2}$

   i Show that $A = \dfrac{\sqrt{3}}{4}x^2 - \dfrac{5}{2}x^2 + 5x$.

   ii Find $\dfrac{dA}{dx}$ and hence find the value of x for which A has a stationary value.

   iii Find $\dfrac{d^2A}{dx^2}$ and hence determine whether this stationary value is a maximum or a minimum. [AQA (B) Jan 2002]

10 Find an equation of the tangent to the curve $y = x + \dfrac{4}{x^2}$ at $(1, 5)$.

11 A curve has equation $y = k\sqrt{x}$. The gradient of the normal at the point $(4, 2k)$ is $-\frac{2}{3}$. Find the value of k.

12 The diagram shows a sketch of the graph of $y = \dfrac{12}{x}$. The point P has coordinates $(3, 4)$.

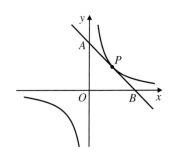

  a The tangent at P crosses the y-axis at A and the x-axis at B.

   i Find the equation of the tangent at P, in the form $px + qy = r$ where p, q and r are integers.

   ii Find the area of triangle AOB.

  b The normal at P crosses the curve again at Q.

   i Show that the equation of the normal at P is $4y = 3x + 7$.

   ii Find the coordinates of Q.

6 Integration

6.1 Integration of rational powers of *x*

Integration of x^n, where $n \neq -1$, and related sums and differences.

In module C1 you integrated polynomials in x of the form ax^n where n is a positive integer. You also evaluated definite integrals and found areas under curves. In C2 the theory is extended to include the case when n is *rational*, with the proviso that $n \neq -1$.

Recall:
C1 Chapter 4.

Note:
The case when $n = -1$ is described in module C3.

The rule is as follows:

$$\int x^n \, dx = \frac{1}{n+1}x^{n+1} + c \quad (n \neq -1)$$

Recall:
Raise the power of x by 1 and divide by the new power.

If *a* is a constant:

$$\int ax^n \, dx = \frac{a}{n+1}x^{n+1} + c \quad (n \neq -1)$$

Tip:
Remember to include the integration constant, c.

Recall also that

$$\int (f(x) \pm g(x)) \, dx = \int f(x) \, dx \pm \int g(x) \, dx$$

Note:
This means that you can integrate term by term.

Example 6.1 Find   **a** $\int 5x\sqrt{x} \, dx$ **b** $\int \sqrt[3]{x} \, dx$.

Step 1: Write in index form.

a $\int 5x\sqrt{x} \, dx = \int 5x^{\frac{3}{2}} \, dx$

Recall:
Laws of indices (Section 1.1).

Step 2: Integrate and simplify if necessary.

$= \frac{5}{\frac{5}{2}}x^{\frac{5}{2}} + c$

$= 2x^{\frac{5}{2}} + c$

Tip:
Take care with the fractions.

b $\int \sqrt[3]{x} \, dx = \int x^{\frac{1}{3}} \, dx$

$= \frac{1}{\frac{4}{3}}x^{\frac{4}{3}} + c$

$= \frac{3}{4}x^{\frac{4}{3}} + c$

Example 6.2 Evaluate $\int_1^2 \frac{3}{x^4} \, dx$.

Recall:
Definite integration (see C1 Section 4.3).

Step 1: Write in index form.

$\int_1^2 \frac{3}{x^4} \, dx = \int_1^2 3x^{-4} \, dx$

Step 2: Integrate.

$= \left[\frac{3}{-3}x^{-3} \right]_1^2$

$= \left[-x^{-3} \right]_1^2$

Step 3: Substitute the limits and evaluate.

$= (-2^{-3}) - (-1^{-3})$

$= -\frac{1}{8} + 1$

$= \frac{7}{8}$

Tip:
Remember to substitute the upper limit first.

Example 6.3 Find $\int \left(x + \dfrac{1}{x}\right)^2 dx$.

Step 1: Expand the function into index form.

$$\left(x + \frac{1}{x}\right)^2 = \left(x + \frac{1}{x}\right)\left(x + \frac{1}{x}\right)$$

$$= x^2 + 1 + 1 + \frac{1}{x^2}$$

$$= x^2 + 2 + x^{-2}$$

Step 2: Integrate term by term.

$$\int \left(x + \frac{1}{x}\right)^2 dx = \int (x^2 + 2 + x^{-2})\, dx$$

$$= \frac{1}{3}x^3 + 2x + \frac{1}{-1}x^{-1} + c$$

$$= \tfrac{1}{3}x^3 + 2x - x^{-1} + c$$

Tip:
You could write this as
$\dfrac{1}{3}x^3 + 2x - \dfrac{1}{x} + c.$

Example 6.4 **a** Write $\dfrac{4x^6 - x}{2x^4}$ in the form $ax^p + bx^q$ where a, b, p and q are rational numbers to be found.

b Hence find $\int \dfrac{4x^6 - x}{2x^4}\, dx$.

Tip:
This is telling you to write the expression in index form before integrating.

Step 1: Write as the sum or difference of terms in the form ax^n.

a $\dfrac{4x^6 - x}{2x^4} = \dfrac{4x^6}{2x^4} - \dfrac{x}{2x^4} = 2x^2 - \tfrac{1}{2}x^{-3}$

Recall:
The laws of indices
(Section 1.1)

Step 2: Compare with the given format.

Hence $a = 2$, $b = -\tfrac{1}{2}$, $p = 2$, $q = -3$.

Step 3: Integrate term by term.

b $\int \dfrac{4x^6 - x}{2x^4}\, dx = \int (2x^2 - \tfrac{1}{2}x^{-3})\, dx$

Tip:
Use your answer from part **a**.

$$= \frac{2}{3}x^3 - \frac{\frac{1}{2}}{-2}x^{-2} + c$$

$$= \tfrac{2}{3}x^3 + \tfrac{1}{4}x^{-2} + c$$

Tip:
Be very careful with negatives and fractions. It is a good idea to write down the working.

Example 6.5 A curve goes through the point $(1, 4)$. The gradient at (x, y) is $4x^{-3}$. Find the equation of the curve.

Step 1: Write the gradient as $\dfrac{dy}{dx}$.

$$\frac{dy}{dx} = 4x^{-3}$$

$$y = \int 4x^{-3}\, dx$$

Recall:
General and particular solutions
(C1 Chapter 4).

Step 2: Integrate with respect to x to find a general solution.

$$= \frac{4}{-2}x^{-2} + c$$

$$= -2x^{-2} + c$$

Tip:
Remember to include the integration constant.

Step 3: Use the condition to find the particular solution.

Since $(1, 4)$ lies on the curve, when $x = 1$, $y = 4$.

Substituting into the equation of the curve gives:

$$4 = -2(1)^{-2} + c \Rightarrow c = 6$$

The equation of the curve is $y = -2x^{-2} + 6$.

Tip:
You could write
$y = 6 - \dfrac{2}{x^2}.$

Example 6.6 Given that $f'(x) = x^2 - \dfrac{1}{x^3}$ and $f(1) = 1$, find $f(x)$.

Step 1: Write in index form.

$$f'(x) = x^2 - \frac{1}{x^3} = x^2 - x^{-3}$$

Step 2: Integrate term by term.

$$f(x) = \int (x^2 - x^{-3})\,dx$$

$$= \frac{1}{3}x^3 - \frac{1}{-2}x^{-2} + c$$

$$= \tfrac{1}{3}x^3 + \tfrac{1}{2}x^{-2} + c$$

Step 3: Use the given condition to find c.

$$f(1) = 1 \implies 1 = \tfrac{1}{3}(1^3) + \tfrac{1}{2}(1^{-2}) + c$$

$$c = 1 - \tfrac{1}{3} - \tfrac{1}{2} = \tfrac{1}{6}$$

$$f(x) = \tfrac{1}{3}x^3 + \tfrac{1}{2}x^{-2} + \tfrac{1}{6}$$

6.2 Area under a curve

When the region is *above* the x-axis:

$$\text{area} = \int_a^b y\,dx$$

Recall:
C1 Section 4.4.

When the region is *below* the x-axis:

$$\text{area} = \left| \int_a^b y\,dx \right|$$

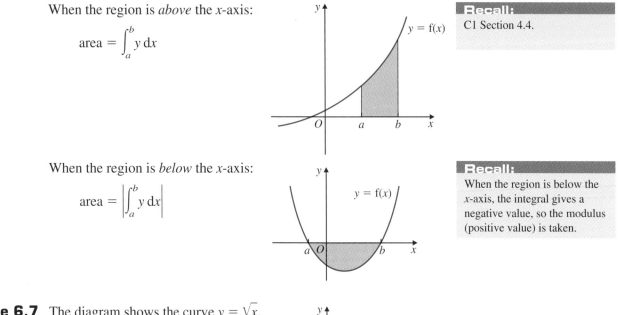

Recall:
When the region is below the x-axis, the integral gives a negative value, so the modulus (positive value) is taken.

Example 6.7 The diagram shows the curve $y = \sqrt{x}$. Calculate the area of the shaded region.

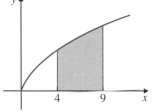

Step 1: Use the formula for the area under a curve.

$$\int_4^9 y\,dx = \int_4^9 \sqrt{x}\,dx$$

$$= \int_4^9 x^{\frac{1}{2}}\,dx$$

Step 2: Integrate.

$$= \left[\frac{1}{\frac{3}{2}} x^{\frac{3}{2}} \right]_4^9$$

Step 3: Substitute the limits and evaluate.

$$= \left[\tfrac{2}{3} x^{\frac{3}{2}} \right]_4^9$$

$$= \tfrac{2}{3}(9^{\frac{3}{2}} - 4^{\frac{3}{2}})$$

$$= \tfrac{2}{3}(27 - 8)$$

$$= 12\tfrac{2}{3}$$

The area is $12\tfrac{2}{3}$.

Example 6.8 The curve $y = \dfrac{1}{x^2}$ and the line $y = x$ intersect at P as shown.

    a Find the coordinates of P. **b** Calculate the area of the shaded region.

Step 1: Solve the simultaneous equations.

a At P, $y = \dfrac{1}{x^2}$ and $y = x$. Solving the equations simultaneously gives

$$x = \frac{1}{x^2}$$

$(\times\, x^2)$ $x^3 = 1$

$$x = 1$$

Substituting into $y = x$ gives $y = 1$, so P is the point $(1, 1)$.

Tip:
You can substitute into either equation.

Step 2: Find the two areas separately.

b Area of region between $y = x$, $x = 1$ and x-axis:

    Area$_1$ = Area of triangle

$$= \frac{1 \times 1}{2}$$

$$= \tfrac{1}{2}$$

Area of region between $y = \dfrac{1}{x^2}$, $x = 1$ and $x = 2$:

$$\text{Area}_2 = \int_1^2 x^{-2}\,dx$$

$$= \left[\frac{1}{-1}x^{-1}\right]_1^2$$

$$= \left[-\frac{1}{x}\right]_1^2$$

$$= -\tfrac{1}{2} - (-1)$$

$$= \tfrac{1}{2}$$

Tip:
You could evaluate $\int_0^1 y\,dx$ where $y = x$.

Tip:
Remember to write terms in index form before integrating.

Step 3: Add the areas.   Total area $= \tfrac{1}{2} + \tfrac{1}{2} = 1$

Example 6.9 The curve $y = 1 + 2\sqrt{x}$ and the line $y = x + 1$ intersect at $P(0, 1)$ and $Q(4, 5)$. Find the area of the region enclosed between the line and the curve, shown shaded in the diagram.

Step 1: Find the area 'under' the curve.

Let the area under the curve be A_1.

$$\text{Area } A_1 = \int_0^4 (1 + 2\sqrt{x})\,dx$$

$$= \int_0^4 (1 + 2x^{\frac{1}{2}})\,dx$$

$$= \left[x + \frac{2}{\frac{3}{2}}x^{\frac{3}{2}}\right]_0^4$$

$$= \left[x + \tfrac{4}{3}x^{\frac{3}{2}}\right]_0^4$$

$$= 4 + \tfrac{32}{3} - 0$$

$$= 14\tfrac{2}{3}$$

Step 2: Find the area 'under' the line.

Let the area under the line be A_2.

$$\text{Area } A_2 = \frac{1 + 5}{2} \times 4 = 12$$

Step 3: Subtract to find the required area.

Required area $= 14\tfrac{2}{3} - 12 = 2\tfrac{2}{3}$

Tip:
Area of a trapezium
$$= \frac{a + b}{2} \times h$$

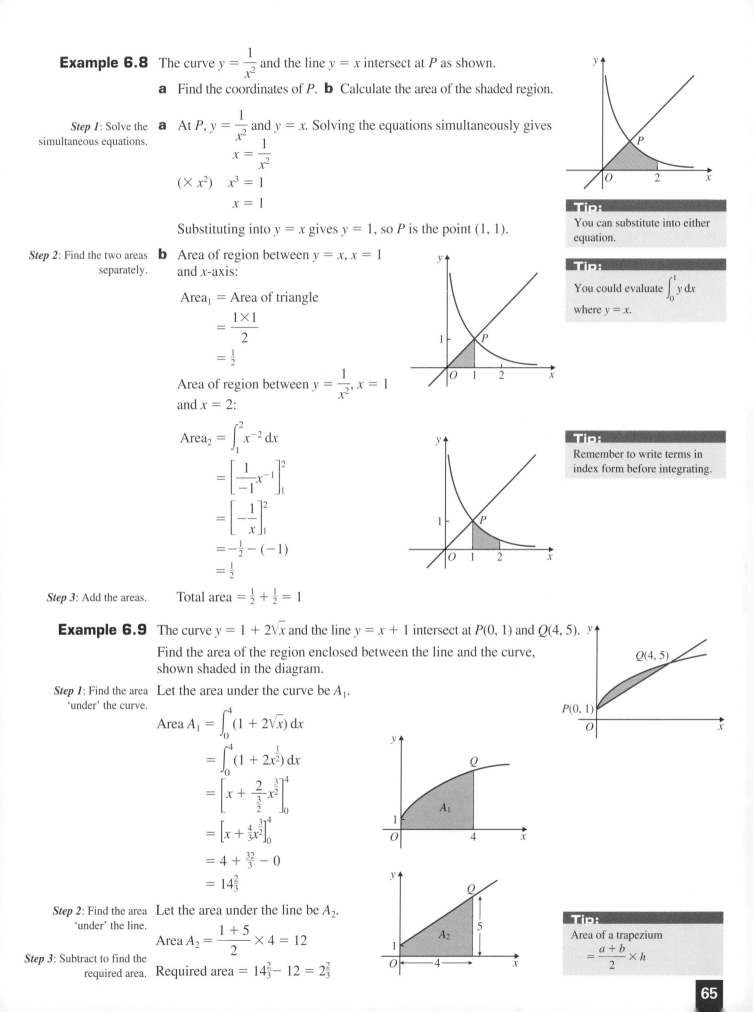

1 Find   **a** $\displaystyle\int \frac{3}{x^2}\,dx$ **b** $\displaystyle\int \frac{3}{\sqrt{x}}\,dx$ **c** $\displaystyle\int \sqrt[4]{x}\,dx.$

2 a Express $x^2\sqrt{x}$ in the form x^k, where k is a rational number.   **b** Find $\displaystyle\int x^2\sqrt{x}\,dx.$

3 Find   **a** $\displaystyle\int (x^2 + \sqrt{x})\,dx$ **b** $\displaystyle\int \frac{x + 2}{\sqrt{x}}\,dx.$

4 Evaluate   **a** $\displaystyle\int_1^2 \left(x - \frac{2}{x^2} \right) dx$ **b** $\displaystyle\int_{-2}^{-1} \frac{1}{x^3}\,dx.$

5 a Write $\dfrac{x^3 + 1}{x^2}$ in the form $x^p + x^q$, where p and q are integers.

b Hence find the value of $\displaystyle\int_1^2 \frac{x^3 + 1}{x^2}\,dx.$

6 A curve passes through the point $(1, 3)$. The gradient at the point (x, y) is $\dfrac{1}{2\sqrt{x}}$.

Find the equation of the curve.

7 The diagram shows the graph of $y = -\dfrac{1}{x^3}$, for $x > 0$.

Find the area of the region enclosed between the curve, the x-axis and the lines $x = 1$ and $x = 2$.

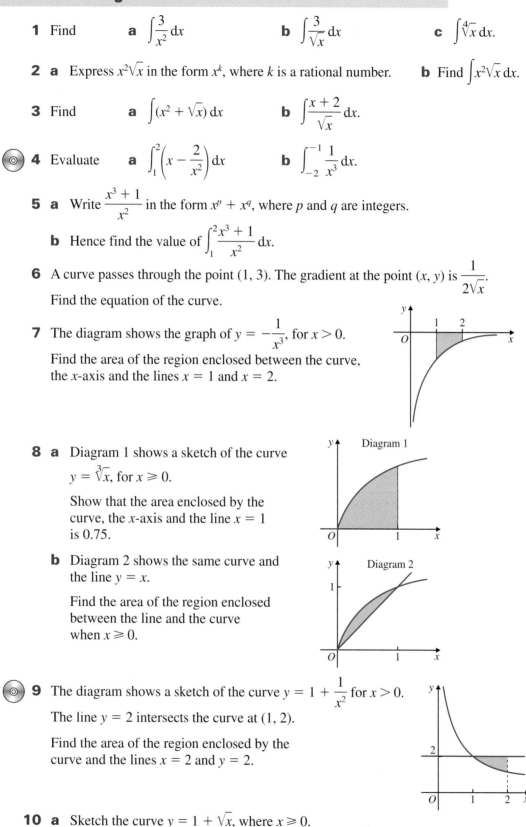

8 a Diagram 1 shows a sketch of the curve

$y = \sqrt[3]{x}$, for $x \geqslant 0$.

Show that the area enclosed by the curve, the x-axis and the line $x = 1$ is 0.75.

b Diagram 2 shows the same curve and the line $y = x$.

Find the area of the region enclosed between the line and the curve when $x \geqslant 0$.

9 The diagram shows a sketch of the curve $y = 1 + \dfrac{1}{x^2}$ for $x > 0$.

The line $y = 2$ intersects the curve at $(1, 2)$.

Find the area of the region enclosed by the curve and the lines $x = 2$ and $y = 2$.

10 a Sketch the curve $y = 1 + \sqrt{x}$, where $x \geqslant 0$.

b Find the area of the region enclosed between the curve and the lines $y = 1$ and $x = 4$.

6.3 Trapezium rule

Approximation of area under a curve using the trapezium rule.

Consider the region enclosed by the curve $y = f(x)$, the x-axis and the lines $x = a$ and $x = b$. To find an approximation of the area of this region, split it into n strips, of equal width h, where $h = \dfrac{b-a}{n}$. Form trapezia by joining the top ends of each strip with a straight line.

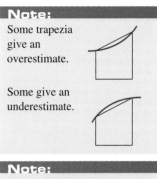

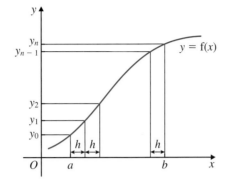

The area under the curve is given by $\displaystyle\int_a^b f(x)\,dx$. An approximate value can be found using the **trapezium rule**:

$$\int_a^b y\,dx \approx \frac{1}{2}h\,[(y_0 + y_n) + 2(y_1 + y_2 + \cdots + y_{n-1})] \text{ where } h = \frac{b-a}{n}.$$

Example 6.10 The diagram shows the region bounded by the curve $y = \sqrt{x-1}$, the x-axis and the lines $x = 2$ and $x = 6$.

a Find an approximation to the area of the region, using the trapezium rule with five ordinates (four strips).

b Is your value an overestimate or an underestimate?

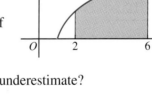

Step 1: Split the region into appropriate strips.

Step 2: Substitute values into the trapezium rule.

a $h = \dfrac{6-2}{4} = 1$, $y_0 = \sqrt{2-1} = \sqrt{1}$, and so on.

x	2	3	4	5	6
y_n	y_0	y_1	y_2	y_3	y_4
$y = \sqrt{x-1}$	$\sqrt{1}$	$\sqrt{2}$	$\sqrt{3}$	$\sqrt{4}$	$\sqrt{5}$

$$\int_a^b y\,dx \approx \frac{1}{2} \times 1[(\sqrt{1} + \sqrt{5}) + 2(\sqrt{2} + \sqrt{3} + \sqrt{4})]$$

$$= 6.764\ldots$$

$$= 6.76 \text{ (3 s.f.)}$$

Step 3: Decide whether the value is an overestimate or underestimate.

b From the graph, it appears that the value calculated by the trapezium rule is an underestimate for the area of the region.

Note on graphical calculators

Some graphical calculators can be programmed to find numerical approximations of definite integrals. These provide a useful check, but you will not be awarded marks for a question on the trapezium rule unless appropriate working has been shown.

Example 6.11 Find an approximation to $\int_0^1 \sqrt{\sin x}\, dx$, where x is in radians, using the trapezium rule with six ordinates (five strips).

Step 1: Split the region into appropriate strips.

$h = \dfrac{1 - 0}{5} = 0.2$

$y_0 = = \sqrt{\sin 0} = 0,\ y_1 = \sqrt{\sin 0.2},\ \dots,\ y_5 = \sqrt{\sin 1}.$

Step 2: Substitute values into the trapezium rule.

x	0	1	2	3	4	5
y_n	y_0	y_1	y_2	y_3	y_4	y_5
$y = \sqrt{\sin x}$	0	$\sqrt{\sin 0.2}$	$\sqrt{\sin 0.4}$	$\sqrt{\sin 0.6}$	$\sqrt{\sin 0.8}$	$\sqrt{\sin 1}$

$$\int_0^1 \sqrt{\sin x}\, dx \approx \tfrac{1}{2} \times 0.2\,[(0 + \sqrt{\sin 1}) + 2(\sqrt{\sin 0.2} + \sqrt{\sin 0.4}$$
$$+ \sqrt{\sin 0.6} + \sqrt{\sin 0.8})]$$
$$= 0.6253\dots$$
$$= 0.625\ (3\ \text{s.f.})$$

Tip:
Remember to set your calculator to radian mode.

SKILLS CHECK 6B: Trapezium rule

1 **a** Sketch the graph of $y = 2^x$.

   b Estimate $\int_0^4 2^x\, dx$, using the trapezium rule with five ordinates (four strips).

   c State whether your estimate is an overestimate or an underestimate.

2 The diagram shows a sketch of $y = \dfrac{1}{1 + x}$ for $x > -1$.

   Estimate $\int_1^2 \dfrac{1}{1 + x}\, dx$ using the trapezium rule with six ordinates (five strips).

3 The following is a set of values, correct to three decimal places, for $y = \cos x$, where x is in radians.

x	0	0.2	0.4	0.6	0.8	1
y	1	0.980	0.921	p	0.697	q

   a Find the value of p and the value of q.

   b Use the trapezium rule and the values of y in the completed table to obtain an estimate for $\int_0^1 \cos x\, dx$.

4 The diagram shows a sketch of $y = \log_{10} x$.

   a Divide the shaded region into five equal-width intervals.

   b Use the trapezium rule to estimate $\int_1^3 \log_{10} x\, dx$.

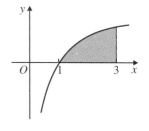

5 **a** Sketch the graph of $y = 10^{-x}$.

   b Estimate $\int_{-2}^{-1} 10^{-x} \, dx$, using the trapezium rule with six ordinates (five strips).

6 The diagram shows the graph of $xy = 12$, for $x > 0$.

Using the trapezium rule with seven ordinates (six strips), estimate the value of $\int_{1}^{7} \dfrac{12}{x} \, dx$.

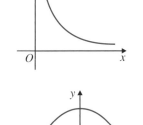

7 The diagram shows a sketch of $y = 4 - x^2$.

   a Estimate the area of the region enclosed between the curve and the x-axis, using the trapezium rule with five ordinates (four strips.)

   b Evaluate the area exactly using integration and calculate the percentage error in taking your answer in **a** as the area.

8 **a** Sketch the curve $y = 3^x + 1$, for values of x between -1 and 3.

   b Use the trapezium rule with five ordinates to estimate the area of the region bounded by the curve, the x-axis and the lines $x = -1$ and $x = 3$.

   c Explain briefly how the trapezium rule could be used to find a more accurate estimate of the area of the required region.

9 The values of a function $f(x)$ are given in the table.

x	1	2	3	4	5	6
$f(x)$	3.5	5	7.5	11	15.5	21

Find an approximate value, using the trapezium rule with six ordinates (five strips), for

   a $\quad \int_{1}^{6} f(x) \, dx$ **b** $\quad \int_{1}^{6} \dfrac{1}{f(x)} \, dx$.

10 **a** Tabulate, correct to two decimal places, the values of the function $f(x) = \dfrac{2}{2 + x^2}$ for values of x from 0 to 2 at intervals of 0.4.

   b Use the values found in part **a** to estimate $\int_{0}^{2} \dfrac{2}{2 + x^2} \, dx$.

SKILLS CHECK **6B EXTRA** is on the CD

Examination practice Integration

1 It is given that $y = x^{\frac{1}{3}}$.

   a Find $\dfrac{dy}{dx}$.

   b **i** Find $\int y \, dx$.

   ii Hence evaluate $\int_{0}^{8} y \, dx$.

[AQA (A) Jan 2002]

2 a Write $x^2\sqrt{x}$ in the form x^k, where k is a fraction.

b The gradient of a curve at the point (x, y) is given by

$$\frac{dy}{dx} = 7x^2\sqrt{x}.$$

Use integration to find the equation of the curve, given that the curve passes through the point $(1, 1)$.

[AQA (B) Jan 2002]

3 Given that $y = x^2 - x^{-2}$, find $\int y\, dx$.

[AQA (A) Jan 2001]

4 a Express $\dfrac{x^5 + 1}{x^2}$ in the form $x^p + x^q$, where p and q are integers.

b Hence, find $\displaystyle\int_1^2 \left(\dfrac{x^5 + 1}{x^2}\right) dx$.

[AQA (B) Nov 2002]

5 The function f is defined for $x \geqslant 0$ by

$$f(x) = x^{\frac{1}{2}} + 2.$$

i Find $\int f(x)\, dx$.

ii Hence show that $\displaystyle\int_0^4 f(x)\, dx = \frac{40}{3}$.

[AQA (A) Jan 2003]

6 A curve has equation $y = x^2 + \dfrac{81}{x^2}$. Its graph is sketched below.

i Find $\displaystyle\int \left(x^2 + \dfrac{81}{x^2}\right) dx$.

ii Hence find the area of the region bounded by the curve, the lines $x = 1$, $x = 3$ and the x-axis.

[AQA (B) Jan 2004]

7 The curve with equation $y = 2x + \dfrac{27}{x^2} - 7$ is defined for $x > 0$, and is sketched opposite.

i Find $\displaystyle\int \left(2x + \dfrac{27}{x^2} - 7\right) dx$.

ii Hence determine the area of the region bounded by the curve, the lines $x = 1$, $x = 2$ and the x-axis.

[AQA (B) Jan 2003]

8 The curve with equation $y = 2x - x^{\frac{3}{2}}$ is defined for $x \geqslant 0$, and is sketched opposite.

Calculate the area of the shaded region bounded by the curve and the x-axis.

[AQA (B) Nov 2003]

9 The diagram shows the graph of

$$y = x^{\frac{3}{2}}, \quad 0 \leqslant x \leqslant 4,$$

and a straight line joining the origin to the point P which has coordinates $(4, 8)$.

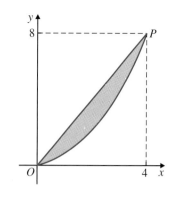

a i Find $\int x^{\frac{3}{2}} \, dx$.

ii Hence find the value of $\displaystyle\int_0^4 x^{\frac{3}{2}} \, dx$.

b Calculate the area of the shaded region.

[AQA (A) June 2003]

10 The diagram shows a sketch of the curve $y = 14 - x^2 - \dfrac{9}{x^2}$ and the line $y = 4$.

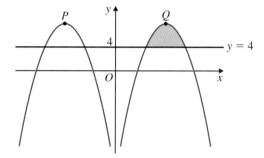

a Find the coordinates of the two stationary points P and Q on the curve.

b i Show that the curve intersects the line $y = 4$ when

$$(x^2 - 9)(x^2 - 1) = 0.$$

ii Hence find the x-coordinates of the four points where the curve intersects the line $y = 4$.

iii Show that the shaded region has area $5\frac{1}{3}$.

[AQA (B) Jan 2002]

11 The diagram shows a sketch of $y = 2^{-x}$.

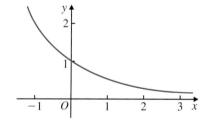

a Use the trapezium rule with six ordinates (five strips) to find an approximation for $\displaystyle\int_0^2 2^{-x} \, dx$.

b By considering the graph of $y = 2^{-x}$, explain with the aid of a diagram whether your approximation will be an overestimate or an underestimate of the true value of $\displaystyle\int_0^2 2^{-x} \, dx$.

12 The diagram shows a sketch of $y = \dfrac{1}{1 + x^2}$.

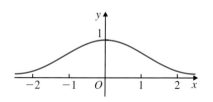

Use the trapezium rule with five ordinates (four strips) to find an approximation for $\displaystyle\int_{-2}^2 \frac{1}{1 + x^2} \, dx$.

Practice exam paper

Answer **all** questions.

Time allowed: 1 hour 30 minutes

A calculator is **allowed** in this paper.

1 The diagram shows a sector of a circle, centre O of radius r cm and angle θ radians.

The length of the chord AB is also r cm.

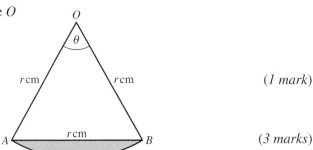

   a Write the value of θ in terms of π.   *(1 mark)*

   b Given that the perimeter of the shaded segment is 25 cm, find the value of r to three significant figures.   *(3 marks)*

2 A sequence of terms is defined by

$$u_{n+1} = ku_n + 3.$$

The first two terms are $u_1 = 15$, $u_2 = 6$.

   a Find the value of the constant k.   *(2 marks)*

   b Given that u_n converges to a limiting value L, find an equation for L and hence find the value of L.   *(3 marks)*

3 The diagrams show two triangles ABC and PQR.

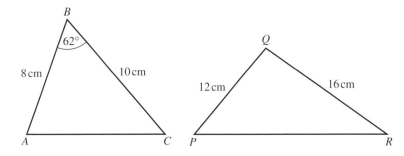

The lengths of PQ and QR are 12 cm and 16 cm respectively. Angle PQR is **obtuse**. The lengths of AB and BC are 8 cm and 10 cm respectively and angle $ABC = 62°$.

   a Calculate the length of AC giving your answer to three significant figures.   *(3 marks)*

   b Given that the area of triangle PQR is the same as the area of triangle ABC, find, to the nearest degree, the size of the **obtuse** angle PQR.   *(4 marks)*

4 **a** Express $(1 + 2Y)^3$ in the form $1 + pY + qY^2 + 8Y^3$, where p and q are integers.   *(3 marks)*

   b Hence expand $(1 - 2\sqrt{x})^3$.   *(2 marks)*

   c It is given that $f(x) = (1 - 2\sqrt{x})^3$.

      i Differentiate $f(x)$ to find $f'(x)$.   *(4 marks)*

      ii Show that $f'(\frac{1}{9}) = -1$.   *(2 marks)*

   d Hence find an equation of the normal to the curve $y = (1 - 2\sqrt{x})^3$ at the point on the curve where $x = \frac{1}{9}$.   *(3 marks)*

5 a Show that the substitution $y = 2^x$ transforms the equation $2^{2x + 2} - 2^{x + 3} + 3 = 0$ into the quadratic equation $4y^2 - 8y + 3 = 0$. *(2 marks)*

b By solving the equation $4y^2 - 8y + 3 = 0$ deduce that one solution of the equation $2^{2x + 2} - 2^{x + 3} + 3 = 0$ is $x = -1$ and find the other solution to four significant figures. *(6 marks)*

6 a Sketch the graph of $y = \tan x$ for $0° \leqslant x \leqslant 360°$ indicating the coordinates of the points where the graph intersects the coordinate axes. *(2 marks)*

b Given that $4 \cos x - 5 \sin x = 0$ find the value of $\tan x$. *(2 marks)*

c Solve the equation

$$\tan 2\theta = 0.8$$

giving all solutions to the nearest $0.1°$ in the interval $0° \leqslant \theta \leqslant 360°$.
No credit will be given for simply reading values from a graph. *(5 marks)*

7 An infinite geometric series has common ratio r.

The first term of the series is 960 and the sum to infinity of the series is 576.

a Show that the value of r is $-\frac{2}{3}$. *(3 marks)*

b Find the sum of the first 12 terms of the series giving your answer to two decimal places. *(3 marks)*

c Given that k is odd find a simplified expression in terms of k for the positive difference between the $(k + 1)$th term and the kth term. *(4 marks)*

8 a i Find $\int \sqrt{x} \, dx$. *(2 marks)*

ii Hence evaluate $\int_1^9 \sqrt{x} \, dx$. *(2 marks)*

b Describe the single transformation by which the curve $y = \sqrt{x}$ can be used to obtain the curve $y = \sqrt{x + 8}$ *(2 marks)*

c The diagram shows the graph of $y = \sqrt{x + 8}$.

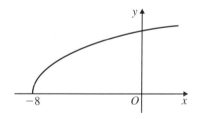

i Use the trapezium rule with five ordinates (four strips) to find an approximation for $\int_{-8}^0 \sqrt{x + 8} \, dx$, giving your answer to three significant figures. *(4 marks)*

ii By considering the graph of $y = \sqrt{x + 8}$ explain whether your approximation will be an overestimate or an underestimate of the true value for $\int_{-8}^0 \sqrt{x + 8} \, dx$. *(2 marks)*

iii Describe the single transformation which can be applied to the graph of $y = \sqrt{x + 8}$ to obtain the graph of $y = \sqrt{4x + 8}$. *(2 marks)*

d Show that $\log_8 \sqrt{4x + 8} = p + q \log_8 (x + 2)$, where p and q are constants to be found. *(4 marks)*

73

Answers

1 a $6x^4y^4$ **b** $16a^8$ **c** $7p^{-1}q^2$
2 a 2   **b** $\frac{1}{9}$ **c** $\frac{1}{3}$ **d** 64   **e** 2
3 $12a^5b^{-\frac{3}{2}}$
4 a $x^{\frac{5}{2}}$ **b** x^{-1} **c** x^5
5 $p^{\frac{1}{3}}$
6 a i 2^{2x} **ii** 2^{3x-3} **b** 2^{5x-3} **c** 3
7 a 3^{4x} **b** $-\frac{1}{3}$ **c** $-\frac{3}{4}$
8 a $y = 3x - \frac{3}{2}$ **b** $4y = 2x + 9$ **c** $x = \frac{3}{2}, y = 3$
9 $x = 2$
10 a 6^{4p} **b** 6^{3q-3} **c** $4p = 3q - 3$ **d** $p = -1, q = -\frac{1}{3}$

1 a i Translation $\begin{bmatrix} 0 \\ 2 \end{bmatrix}$ **ii** Reflection in the x-axis
2 Curve for $y = \cos(x + 30°)$: translate $y = \cos x$, $30°$ to the left.
3 a Translation $\begin{bmatrix} 0 \\ 1 \end{bmatrix}$ **b** Stretch in the y-direction, factor 4
   c Reflection in the x-axis   **d** Stretch in the x-direction, factor $\frac{1}{2}$
4 a $(1, -3)$
   b i Stretch in the y-direction, factor 2
      ii Curve goes through $(0, 0)$, $(1, 6)$, $(2, 0)$; $A(1, 6)$
5 Curve for $\sin(x - \frac{1}{4}\pi)$: translate $y = \sin x$, $\frac{1}{4}\pi$ to the right
6 a i Stretch in the y-direction, factor 4   **ii** Stretch in the x-direction, factor $\frac{1}{2}$
   b $y = 4x^2$; the curves are the same
7 a Graph through $(-2, 0)$, $(-1, -0.5)$, $(0, 0)$, $(2, 1)$
   b Graph through $(-2, -1)$, $(0, 0)$, $(1, 0.5)$, $(2, 0)$
   c Graph through $(-2, 0)$, $(0, -1)$, $(1, -1.5)$, $(2, -1)$
8 a Similar graphs to Example 1.7, with $y = 2^x$ going through $(0, 1)$;
      $y = 2^x + 2$ goes through $(0, 3)$.
   b Reflection of $y = 2^x$ in the x-axis   **c** $y = 2^{x-2}$ goes through $(0, \frac{1}{4})$
9 b, d
10 Stretch in the x-direction, factor $\frac{1}{2}$. Asymptotes at $x = 45°$, $x = 135°$;
   goes through $(0°, 0°)$, $(90°, 0°)$, $(180°, 0°)$

1 a i 3^{-3} **ii** 3^{2x} **b** -4
2 a i $3^{\frac{1}{2}}$ **ii** $3^{x-\frac{1}{2}}$ **b** $-\frac{1}{2}$
3 i $2^{\frac{1}{2}}$ **ii** $2^{\frac{5}{2}}$ **iii** $-\frac{1}{2}$
4 a x^{-1} **b** x^4 **c** $x^{\frac{3}{4}}$
5 $16x^2$
6 b Curve for $y = f(x - \frac{1}{2}\pi)$: translate $\frac{1}{2}\pi$ to the right
7 i Stretch in the x-direction, factor $\frac{1}{2}$ **ii** $x = 90°$ ($x = \frac{1}{2}\pi$ in radians)
8 a $\begin{bmatrix} 0 \\ 2 \end{bmatrix}$ **b** Stretch in the x-direction, factor $\frac{1}{100}$
9 a Translation $\begin{bmatrix} 1 \\ 0 \end{bmatrix}$ **b** Stretch in the y-direction, factor $\frac{1}{2}$
10 a $A(-4, 0)$, $B(-2, 4)$, $C(0, 2)$, $D(2, 0)$
   b $A(-8, 0)$, $B(-4, 2)$, $C(0, 1)$, $D(4, 0)$

1 0, 3, 8, 15
2 a 1.3333..., 1.5555..., 1.7037..., 1.8024...,   **b** 2
3 a 4, 8, 16, 32   **b** 7, 15, 31, 63
4 a $x_{n+1} = 0.9x_n$ **b** £5900
5 10, 3.1623, 1.7783, 1.3335, 1.1548, 1.0746, 1.0366
6 a $-4.833...$, $-5.034...$, $-4.993...$, $-5.001...$, $-4.999...$, $-5.000...$
   b -5 **c** $L = -5$
7 a 75   **b** 31
8 a 42   **b** 15
9 $-4, -9, -7$
10 a 40   **b** $2\frac{103}{210}$

1 a 5, 1190   **b** $-3, -610$
2 $u_n = n - \frac{1}{2}, S_{200} = 20\,000$
3 a 3   **b** $3n + 5$ **c** 1100
4 a 3   **b** $a = 9, d = -1$ **c** -165
5 a £50   **b** £10 500
6 a 741   **b** 1179
7 99
8 9
9 a 8, 10, 12   **b** 2   **c** 20   **d** 540
10 a 15 months   **b** £725

1 a i 31 250   **ii** 39 062
   b i $\frac{7}{64}$ **ii** $13\frac{57}{64}$ **iii** 14
   c i 64   **ii** 43
2 $-59\,048$
3 £123 000
4 a $a = 40, r = \frac{1}{4}$ **b** $53\frac{1}{3}$
5 a $-\frac{1}{3}$ **b** $\frac{1}{324}$
6 a 699 048   **b** 3
7 137 262
8 a £512   **b** £1023
9 a 16   **b** $p = 14, q = 6$
10 1

1 $16 - 96x + 216x^2 - 216x^3 + 81x^4$
2 $1 + 28y + 336y^2 + 2240y^3$
3 a $243 - 810x + 1080x^2$ **b** $A = 1215, B = -3564, C = 3780$
4 a $1 + 12x + 60x^2 + 160x^3$ **b** 1.1262 (4 d.p.)
5 0.972 333 8 (7 d.p.)
6 $40\sqrt{6}$
7 4
8 a $n = 8, k = -\frac{1}{2}$ **b** $-7x^3$
9 a $1 + 6ax + 15a^2x^2$ **b** $a = 3, b = 2$
10 a $1 - 15x + 90x^2 - 270x^3$ **b** -180

1 a $a = 8, d = 4$ **b** 5300
2 b $a = -13, d = 3$
3 a 4   **b** 5250   **c** 250
4 a 392   **b i** 47, 44, 41, 38   **ii** Show $u_{16} > 0, u_{17} < 0$
5 a 10.5, 11   **b** 0.5   **c** 30   **d** 532.5
6 a $u_n = 12 + 3n$ **c** 16 miles
7 b 48   **c** 96
8 b $10(0.9)^{n-1}$ **d** 100
9 b i 100   **ii** 368.928
10 a $4\sqrt{2}$ **b** $8(\sqrt{2} + 1), k = 8$
11 a ii $b = 2000\left(1 + \dfrac{p}{100}\right), c = 2000\left(1 + \dfrac{p}{100}\right)^2$
   b ii $u_n = 2000(1.08)^n$ **iii** £4317.85
12 a $1200r, 1200r^2$ **b ii** £7500
13 a 75 150   **b** $S_n = 3^n - 1$
14 $x^5 + 10x^4 + 40x^3 + 80x^2 + 80x + 32$
15 a $1 + 8x + 28x^2 + 56x^3$ **b** 112
16 $x^4 - 9x^3 + 21x^2 - 35x + 15$

1 7.8 cm
2 15.8 cm (3 s.f.)
3 a $24.5°$ (3 s.f.)   **b** $11.0\,\text{cm}^2$ (3 s.f.)
4 a $52.1°$ (3 s.f.)   **b** $92.9°$ (3 s.f.)
5 a 8.25 cm (3 s.f.)   **b** $35.8\,\text{cm}^2$ (3 s.f.)
6 b i $11.6\,\text{cm}^2$ (3 s.f.), $17.7\,\text{cm}^2$ (3 s.f.)
      ii 4.10 cm (3 s.f.), 9.19 cm (3 s.f.)
7 $55.9°$ (3 s.f.)
8 a 19.2 cm (3 s.f.)   **b** $69.2\,\text{cm}^2$ (3 s.f.)

9 a 17.9 m (3 s.f.)   **b** 57.1° (3 s.f.)
c 60.9° (3 s.f.)   **d** 265 m² (3 s.f.)
10 a 30°, 6.46 cm (3 s.f.)   **b** 19.3 cm (3 s.f.)
c $\sin A = 0.5 \Rightarrow a = 30°$, 150° giving two triangles with same base and same perpendicular height (see CD for diagrams).

SKILLS CHECK 3B (page 35)

1 a 4.89 radians (3 s.f.)   **b** 85.9° (3 s.f.)
2 a 120°   **b** 135°   **c** 270°   **d** 105°
3 a $\frac{1}{4}\pi$ **b** $\frac{5}{6}\pi$ **c** $\frac{11}{6}\pi$ **d** $\frac{4}{3}\pi$
4 a 3 cm   **b** 13 cm
c 7.06 cm² (3 s.f.)   **d** 7.5 cm²
5 a 1.2 radians   **b** 64.9 cm² (3 s.f.)
6 a 1.5 radians   **b** 21 cm
7 a 0.848 radians   **b** 27.1 cm² (3 s.f.)   **c** 3.14 cm² (3 s.f.)
8 a $6 + 17\theta$ **c** 12.75 cm²
9 a 4.84 cm (3 s.f.)   **b** 1.05 cm (3 s.f.)   **c** 16.9 cm (3 s.f.)
d 13.5 cm²   **e** 1.05 cm² (3 s.f.)   **f** 12.5 cm² (3 s.f.)
10 a 4.57 cm (3 s.f.)   **b** 1.34 cm² (3 s.f.)

SKILLS CHECK 3C (page 43)

1 a 17°, 163° (nearest °)   **b** 60°, 300°
c 124°, 304° (nearest °)   **d** 10°, 110°, 130°, 230°, 250°, 350°
e 105°, 165°, 285°, 345°   **f** 90°
2 a $\frac{1}{3}\pi, \frac{2}{3}\pi$ **b** $\frac{3}{4}\pi$ **c** $\frac{2}{3}\pi$
d $\frac{7}{12}\pi, \frac{11}{12}\pi$ **e** $\frac{1}{9}\pi, \frac{5}{9}\pi, \frac{7}{9}\pi$ **f** $\frac{1}{24}\pi, \frac{7}{24}\pi, \frac{13}{24}\pi, \frac{19}{24}\pi$
3 ± 0.72 radians, ± 5.56 radians (2 d.p.)
4 $-90°, -30°, 30°, 90°$
5 $-360°, -315°, -225°, -180°, 0°, 45°, 135°, 180°, 360°$
7 $\frac{4}{3}$
8 a 1   **b** $\frac{1}{12}\pi, \frac{5}{12}\pi, \frac{3}{4}\pi$ [0.26ᶜ (2 d.p.), 1.31ᶜ (2 d.p.), 2.36ᶜ (2 d.p.)]
9 $\frac{1}{2}\pi, \frac{11}{6}\pi$ [1.57ᶜ (2 d.p.), 5.76ᶜ (2 d.p.)]
10 b 0°, 120°, 240°, 360°   **c** 0°, 60°, 120°, 180°

Exam practice 3 (page 43)

1 a 77 m (nearest m)   **b** 874 m² (3 s.f.)
2 16 hectares
3 a 60 cm   **b** 98.6 cm (nearest mm)
5 a 4.5 cm   **b** 6.75 cm²
6 a i 3 cm, $3\sqrt{3}$ cm   **ii** 2π cm   **iii** 6π cm²   **b** $m = 3$
7 10°, 130°
8 1.57 (2 d.p.), 3.67 (2 d.p.)
9 2.09ᶜ (2 d.p.), 5.24ᶜ (2 d.p.)
10 a $\frac{3}{4}$ **b** 7.4° (1 d.p.), 43.4° (1 d.p.), 79.4° (1 d.p.)
11 b $0, \pi, \frac{7}{6}\pi, \frac{11}{6}\pi$ [0ᶜ, 3.14ᶜ (2 d.p.), 3.67ᶜ (2 d.p.), 5.76ᶜ (2 d.p.)]
12 a i $1 - \cos^2 x$ **b** $-2, \frac{1}{4}$ **c** 75.5° (1 d.p.), 284.5° (1 d.p.)
13 c i $-2, \frac{1}{2}$ **ii** $\frac{1}{6}\pi, \frac{5}{6}\pi$ [0.52ᶜ (2 d.p.), 2.62ᶜ (2 d.p.)]
14 121.5° (1 d.p.), 278.5° (1 d.p.)
15 a ii $\frac{5}{12}$ **b** 0.395 radians (3 s.f.)   **c ii** 32 cm²
16 a i $\frac{1}{3}\pi$ **ii** 2π cm   **b i** 19 cm   **ii** 25 cm² (2 s.f.)

SKILLS CHECK 4A (page 50)

1 a 3   **b** 2   **c** 3   **d** $\frac{1}{2}$
2 a 9   **b** -2 **c** $\frac{3}{2}$ **d** $\frac{3}{2}$
3 a 2   **b** 36
4 a 5   **b** $-\frac{1}{2}$ **c** 3
5 $\log_2 p + \frac{1}{2}\log_2 q - \frac{1}{2} - \frac{3}{2}\log_2 r$
6 36
7 a i 1   **ii** 3   **b** -3
8 a 4.75 (3 s.f.)   **b** 0.945 (3 s.f.)   **c** 2.92 (3 s.f.)
9 a $a = d = \log_a b$ **b** 55
10 b $x = 0, 2$

Exam practice 4 (page 51)

1 a 3   **b** $2\log_2 3$ **c** $3 + 2\log_2 3$
2 b i $\frac{1}{1}$ **ii** $\sqrt{2}$
3 b $3\sqrt{3}$
4 b i 12   **ii** -1.5
5 a i 1   **ii** 3   **b** -3
6 a 3   **b** 3
7 3.73 (2 d.p.)
8 7.1 (1 d.p.)
9 b $x = \frac{1}{2}$
10 a 2^{4x+2}

SKILLS CHECK 5A (page 54)

1 a $-3x^{-4}$ **b** $-10x^{-6}$ **c** $10x^{\frac{3}{2}}$ **d** $-3x^{-\frac{3}{2}}$
2 $-2x^{-3}$
3 a $\frac{1}{4}x^{-\frac{3}{2}}$ **b** $\frac{1}{4}$
4 5
5 $\frac{3}{8}, 3, -\frac{1}{9}$
6 ± 2
7 $\pm\frac{1}{2}\sqrt{2}$
8 -4
10 42

SKILLS CHECK 5B (page 58)

1 a $3x^{\frac{1}{2}} - 10x^{\frac{3}{2}} + 2$ **c** $x - 5y - 1 = 0$
2 $x - 2y + 9 = 0$
3 $y = 5x + 80$
4 a $y = -7x - 10$ **b** $\frac{3}{4}$, minimum
5 a 2   **b** $8\sqrt{2}$
6 b 9   **c** 972 cm²
7 b $x + 48y - 32 = 0$
8 c 1728 cm²
9 a $\frac{3}{4}x^{-\frac{3}{2}}$ **b** $-10\frac{2}{3}$
10 $9x - 2y - 11 = 0$

Exam practice 5 (page 59)

1 a $x^{\frac{5}{2}}$ **b** 67.5
2 a $1 - 8x^{-3}$ **b** (2, 3)   **c** 1.5, minimum
3 i $2 - 54x^{-3}$ **ii** 3
4 i $\frac{1}{2\sqrt{x}}$ **ii** $\frac{1}{4}$
5 i $2 - \frac{3}{2}\sqrt{x}$
6 iii $-\frac{1}{64}$, maximum
7 i $2x - \frac{162}{x^3}$ **iii** ± 3 **iv** 18
8 a i $4\pi x - \frac{1000}{x^2}$ **iii** $4\pi + \frac{2000}{x^3}$ **v** 4.3 (1 d.p.), minimum
b 349 cm²
9 b ii $\frac{\sqrt{3}}{2}x - 5x + 5$, 1.21 (3 s.f.)   **iii** $\frac{\sqrt{3}}{2} - 5$, maximum
10 $y + 7x = 12$
11 6
12 a i $3y + 4x = 24$ **ii** 24 units²   **b ii** $\left(-\frac{16}{3}, -\frac{9}{4}\right)$

SKILLS CHECK 6A (page 66)

1 a $-\frac{3}{x} + c$ **b** $6\sqrt{x} + c$ **c** $\frac{4}{5}x^{\frac{5}{4}} + c$
2 a $x^{\frac{5}{2}}$ **b** $\frac{2}{7}x^{\frac{7}{2}} + c$
3 a $\frac{1}{3}x^3 + \frac{2}{3}x^{\frac{3}{2}} + c$ **b** $\frac{2}{3}x^{\frac{3}{2}} + 4x^{\frac{1}{2}} + c$

4 a $\frac{1}{2}$ **b** $-\frac{3}{8}$
5 a $x + x^{-2}$ **b** 2
6 $y = \sqrt{x} + 2$
7 $\frac{3}{8}$
8 b 0.25
9 $\frac{1}{2}$
10 b $5\frac{1}{3}$

SKILLS CHECK 6B (page 68)

1 b 22.5   **c** Underestimate
2 0.406 (3 s.f.)
3 a $p = 0.825$, $q = 0.540$ **b** 0.84 (2 d.p.)
4 a Strips have width 0.4   **b** 0.559 (3 s.f.)
5 b 39.8 (3 s.f.)
6 24.3 (3 s.f.)
7 a 10   **b** $10\frac{2}{3}$, 6.25%
8 b $30\frac{2}{3}$ **c** Split into more strips
9 a 51.25   **b** 0.655 (3 s.f.)
10 a 1, 0.93, 0.76, 0.58, 0.44, 0.33   **b** 1.35 (3 s.f.)

Exam practice 6 (page 69)

1 a $\frac{1}{3}x^{-\frac{2}{3}}$ **b i** $\frac{3}{4}x^{\frac{4}{3}} + c$ **ii** 12
2 a $x^{\frac{5}{2}}$ **b** $y = 2x^{\frac{7}{2}} - 1$
3 $\frac{1}{3}x^3 + x^{-1} + c$
4 a $x^3 + x^{-2}$ **b** $4\frac{1}{4}$
5 a i $\frac{2}{3}x^{\frac{3}{2}} + 2x + c$
6 i $\frac{1}{3}x^3 - 81x^{-1} + c$ **ii** $62\frac{2}{3}$
7 i $x^2 - \dfrac{27}{x} - 7x + c$ **ii** 9.5
8 3.2
9 a i $\frac{2}{5}x^{\frac{5}{2}} + c$ **ii** 12.8   **b** 3.2

10 a $(\sqrt{3}, 8), (-\sqrt{3}, 8)$ **b ii** $-3, 3, -1, 1$
11 a 1.09 (3 s.f.)   **b** Overestimate
12 2.2

Practice exam paper (page 72)

1 a $\dfrac{\pi}{3}$ **b** $r = 12.2$ to 3 s.f.
2 a $k = 0.2$ **b** $L = kL + 3$, $L = 3.75$
3 a $AC = 9.43$ cm to 3 s.f.   **b** 158° (to the nearest degree)
4 a $1 + 6Y + 12Y^2 + 8Y^3$ **b** $1 - 6\sqrt{x} + 12x - 8x\sqrt{x}$
    c i $-\dfrac{3}{\sqrt{x}} + 12 - 12\sqrt{x}$ **d** $y - \frac{1}{27} = 1(x - \frac{1}{9})$
5 b 0.5850 to 4 s.f.
6 a

b $\tan x = \frac{4}{5}$
c To the nearest 0.1°, $\theta = 19.3°, 109.3°, 199.3°, 289.3°$
7 b 571.56 to 2 d.p.   **c** $2400(\frac{2}{3})^k$
8 a i $\frac{2}{3}x^{\frac{3}{2}} + c$ **ii** $17\frac{1}{3}$
    b Translation $\begin{bmatrix} -8 \\ 0 \end{bmatrix}$
    c i 14.6 to 3 s.f.   **ii** Underestimate
      iii Stretch in the x-direction, scale factor $\frac{1}{4}$.
    d $p = \frac{1}{3}$ and $q = \frac{1}{2}$

SINGLE USER LICENCE AGREEMENT FOR CORE 2 FOR AQA CD-ROM
IMPORTANT: READ CAREFULLY

WARNING: BY OPENING THE PACKAGE YOU AGREE TO BE BOUND BY THE TERMS OF THE LICENCE AGREEMENT BELOW.

This is a legally binding agreement between You (the user or purchaser) and Pearson Education Limited. By retaining this licence, any software media or accompanying written materials or carrying out any of the permitted activities You agree to be bound by the terms of the licence agreement below.

If You do not agree to these terms then promptly return the entire publication (this licence and all software, written materials, packaging and any other components received with it) with Your sales receipt to Your supplier for a full refund.

YOU ARE PERMITTED TO:

- Use (load into temporary memory or permanent storage) a single copy of the software on only one computer at a time. If this computer is linked to a network then the software may only be used in a manner such that it is not accessible to other machines on the network.

- Transfer the software from one computer to another provided that you only use it on one computer at a time.

- Print a single copy of any PDF file from the CD-ROM for the sole use of the user.

YOU MAY NOT:

- Rent or lease the software or any part of the publication.

- Copy any part of the documentation, except where specifically indicated otherwise.

- Make copies of the software, other than for backup purposes.

- Reverse engineer, decompile or disassemble the software.

- Use the software on more than one computer at a time.

- Install the software on any networked computer in a way that could allow access to it from more than one machine on the network.

- Use the software in any way not specified above without the prior written consent of Pearson Education Limited.

- Print off multiple copies of any PDF file.

ONE COPY ONLY

This licence is for a single user copy of the software

PEARSON EDUCATION LIMITED RESERVES THE RIGHT TO TERMINATE THIS LICENCE BY WRITTEN NOTICE AND TO TAKE ACTION TO RECOVER ANY DAMAGES SUFFERED BY PEARSON EDUCATION LIMITED IF YOU BREACH ANY PROVISION OF THIS AGREEMENT.

Pearson Education Limited and/or its licensors own the software.
You only own the disk on which the software is supplied.

Pearson Education Limited warrants that the diskette or CD-ROM on which the software is supplied is free from defects in materials and workmanship under normal use for ninety (90) days from the date You receive it. This warranty is limited to You and is not transferable. Pearson Education Limited does not warrant that the functions of the software meet Your requirements or that the media is compatible with any computer system on which it is used or that the operation of the software will be unlimited or error free.

You assume responsibility for selecting the software to achieve Your intended results and for the installation of, the use of and the results obtained from the software. The entire liability of Pearson Education Limited and its suppliers and your only remedy shall be replacement free of charge of the components that do not meet this warranty.

This limited warranty is void if any damage has resulted from accident, abuse, misapplication, service or modification by someone other than Pearson Education Limited. In no event shall Pearson Education Limited or its suppliers be liable for any damages whatsoever arising out of installation of the software, even if advised of the possibility of such damages. Pearson Education Limited will not be liable for any loss or damage of any nature suffered by any party as a result of reliance upon or reproduction of or any errors in the content of the publication.

Pearson Education Limited does not limit its liability for death or personal injury caused by its negligence.

This licence agreement shall be governed by and interpreted and construed in accordance with English law.